IS SOCIALISM DOOMED?

IS SOCIALISM DOOMED?
The Meaning of Mitterrand

Daniel Singer

New York   Oxford
OXFORD UNIVERSITY PRESS
1988

Oxford University Press

Oxford New York Toronto
Delhi Bombay Calcutta Madras Karachi
Petaling Jaya Singapore Hong Kong Tokyo
Nairobi Dar es Salaam Cape Town
Melbourne Auckland

and associated companies in
Berlin Ibadan

Copyright © 1988 by Oxford University Press, Inc.

Published by Oxford University Press, Inc.,
200 Madison Avenue, New York, New York 10016

Oxford is a registered trademark of Oxford University Press

All rights reserved. No part of this publication may be reproduced,
stored in a retrieval system, or transmitted, in any form or by any means,
electronic, mechanical, photocopying, recording, or otherwise,
without the prior permission of Oxford University Press.

Library of Congress Cataloging-in-Publication Data
Singer, Daniel, 1926–
Is socialism doomed?
Bibliography: p.
Includes index.
1. France—Politics and government—1981–   .
2. France—Economic conditions—1945–   . 3. France—
Economic policy—20th century. 4. Socialism—France—
History—20th century. 5. Mitterrand, François, 1916–
Influence. I. Title.
DC423.S565 1988   944.083'8   87-20356
ISBN 0-19-504925-X

2 4 6 8 9 7 5 3 1
Printed in the United States of America
on acid-free paper

Contents

Introduction: Is the Future American? 3

Part I The Shifting Context 13

1. Changing France in an Expanding Europe 15
2. The Left Bewitched and Bewildered 37
3. Seeds of Defeat in Victory 56
4. One Character in Search of a Role 78

Part II The Road to Surrender 95

5. The Fall from Grace 97
6. The Retreat 129
7. Cultural Counterrevolution 153
8. The Conversion 189

Part III The European Dimension 219

9. Reagan's Best Ally 223
10. In the Footsteps of Predecessors 232
11. Europe and the Nation-State 242
12. Socialism and National Frontiers 251

Part IV Lessons from France 257

13. The Case For and Against Mitterrand 259
14. Crisis and Polarization,
    or For Whom the Bell Tolls 269
15. Socialism to Be "Reinvented" 277
16. The Negative Lesson 286

Notes 297
Index 315

IS SOCIALISM DOOMED?

Introduction:
Is the Future American?

> Men at some time are masters of their fates
> Shakespeare, *Julius Caesar*

The rule of the Left in France ended without drama. If T. S. Eliot's metaphor were not so worn, one might add "with barely a whimper." The Socialists just tiptoed off the governmental stage. Their reign had begun on May 10, 1981, when a half-surprised world learned about the victory of the Left in France and the election of sixty-four-year-old François Mitterrand to the presidency of the Republic. It came to a stop on March 16, 1986, with the victory of the Right and the return to parliamentary power of Jacques Chirac, Raymond Barre, and Valéry Giscard d'Estaing, the men of yesterday. True, it was not a clean break. Mitterrand's presidential mandate was for seven years and thus still had two years to run. The Fifth Republic, established in 1958 by General Charles de Gaulle, found itself in an unprecedented situation, with a conservative government supposed to govern and a Socialist president presumed to preside. In any case, the supremacy of the Left and, with it, the socialist experiment as such were well over.

Both ended without drama and without real passion, if one excepts the frustration of people losing their jobs and the satisfaction of those helping to fill them. Admittedly, the transfer of power was expected and discounted. President Mitterrand had altered the electoral law to mandate proportional representation in elections to the National Assembly in order to prevent the victory of the Right at the polls from turning into a landslide, and his calculation proved correct. Besides, the Socialists, although deprived of office, had the consolation of remaining the largest party in France and entertained the hope of regaining power soon. If the Left did not cry too much, the respectable Right—that is to say, the conservatives who refused to mix, at least in public, with Jean-Marie Le Pen and his xenophobic National Front—did not rejoice too loudly. Its victory in terms of seats in the National Assembly was much lower than expected, not because of any sudden recovery of the Left but because of the grimly successful performance of the National Front. The only people with valid reasons to celebrate, to drink champagne that night, were Le Pen and his accomplices. For the first time since the rise of Adolf Hitler, a party of the extreme Right had received one tenth of the vote in a major election.[1]

The absence of a surprise and of a clean break or landslide cannot, on their own, explain the lack of political passion. There are deeper reasons for this. People may have been reticent on this occasion because the election did not matter as much as elections had in the past. Although significant, the difference between the respectable Right and a *gauche respectueuse*[2]—respectful, that is, of the established order—was not crucial. The French were being offered a marginal choice within the consensus, as are the Americans, the British, and the Germans, and that was the novelty. Hitherto, whereas elections in other nations involved merely a change of government or administration, it always seemed as though the French regime itself were at stake. Rightly or wrongly, it was always feared—or hoped—that the Left in office, particularly if it included the Communists, might not respect the rules of the game or might even threaten the very foundations of society. And then the Left came, won, but did not conquer. Hell did not break loose. In five years, the Socialists destroyed the myth. To appreciate properly the indifference with which the departure of the Left was greeted, one must recall the enthusiasm on the day of its arrival.

Just before 8:00 P.M. on that fateful May 10, a Sunday, the whole nation was tense. Since polling booths in France do not close at the same time, as they do in the United States, the results cannot be announced until voting has ceased throughout the country. Thus, although pollsters and politicians already knew the verdict, the general public did not. Millions were glued to their television sets. The chief newscasters, darlings of the regime and therefore likely to sink together with Giscard, the outgoing president (at the time, all television in France was still state owned and controlled), looked grim, and this was being interpreted by left-wingers as a good omen. An indication, however, is not a proof, and after so many illusions, so many disappointments, they kept their fingers crossed. And then, as the clock struck eight, the three television channels confirmed their hopes: Mitterrand, not by knockout, but clearly on points. Later that night, official results were announced: In metropolitan France, Mitterrand had gathered 15.6 million votes, or 52.2 percent of the total; Giscard, the remaining 14.2 million.

For the first time in French history, a Socialist was elected president of the Republic through universal suffrage. The Left was taking over after twenty-three years of right-wing rule—and a nominally united Left at that, which would soon mean the presence of Communist ministers in a Western European government for the first time since the advent of the cold war. This was enough to cause more than misgivings in many a Western chancery, apprehension in Bonn, hardly concealed displeasure in Washington, and enough to provoke surprise, dismay, and rage among France's moneyed masters and their hangers-on. It was also enough to set off an explosion of joy among millions hitherto convinced that they were, in practical terms, disfranchised.

From the working-class suburbs and poorer districts they flocked to the center; in Paris it was to the place de la Bastille, where the notorious prison fortress once stood. Generations mingled. The elder, those who remembered May 1958, when a *Putsch* by colonial settlers and army commanders in Algeria had brought the Fourth Republic to its knees and General de Gaulle back to power, had reasons to believe that the rules of the game were so twisted as to keep the Right permanently in power. Joining them were the veterans of the great revolt of students and workers in May 1968. Their feelings were mixed: Although their side had won, this was not quite their victory, not the one they had imagined. Most numerous, however, were the youngsters, almost discovering politics with this electoral victory of the Left.

They had not come to the Bastille to storm any fortress, real or metaphorical. They did not expect full employment for breakfast and a socialist paradise the day after. They were not awaiting a magic dawn. Of the slogans the two left-wing parties had once put on their programs, the Socialist "change life" was too sanguine for the occasion; the Communist proposal to "change course" seemed more appropriate. Yet even this was exhilarating. To make a fresh start in a new direction, inspired by the vision of a different future, was in itself intoxicating. They had come to dance and celebrate. The crowd was generous, strangely unrevengeful and good-humored. It applauded the Socialists and Communists who gave impromptu speeches. It mocked the absent losers rather gently. Even the sudden storm at midnight, with rain pouring down on the dancers, could not disturb their joyful mood. After five years of Socialist practice, even this celebration, this unrevolutionary fiesta, looks rather strangely distant and naïvely utopian.

Is socialism, then, a figment of the imagination, an inspiring vision unconnected with the rugged reality of modern society? For years, as the Soviet dream was turning into a nightmare, it was still possible to argue that socialism, designed for the advanced countries of Western Europe, was doomed to wither in the arid soil of backward Mother Russia. The same excuse cannot be used to explain the failure of the French experiment. To say that the French case proves nothing, since no genuine attempt was made by Mitterrand to turn France into a socialist country or even to begin moving in that direction, is not enough. If not, why not? France, after all, with its radical tradition and its revolutionary memories, is different from Britain or Germany. The French Left, while in opposition, did promise to make "a break with capitalism," and on reaching office, it did nationalize a number of industrial conglomerates and merchant banks. Yet within months—two years at the utmost—it was entirely converted, extolling private enterprise and singing the praises of profit. Why this huge gap between promise and fulfillment? What are the reasons for the utter ideological and political bankruptcy of the French Left?

The questions at the heart of this book are not of interest just to the French or to Socialists. With the Labour party apparently unable to return to office in Britain and the Social Democrats pondering how they will get back into government in Germany, the role of the Left in the new historical period opens up a much wider debate. Is it still possible to believe that the countries of Western Europe will invent a social system different from the American model and find in it an incentive to unite, as well as an instrument for resisting the pressure of the United States? Can they forge a radical alternative likely to act as a magnet for the other half of Europe, ultimately including the Soviet Union? These are problems not irrelevant to those in the White House trying to make up their minds how best to handle the European Economic Community—as a junior partner or an economic rival?—or to those in the Kremlin beginning to wonder whether the search for an understanding with the other nuclear giant is the only possible strategy. Yet these questions also go well beyond the traditional calculations of politicians.

With its sudden swings and spectacular volte-faces, France is a striking example of the two contrasting trends affecting Western Europe and the two conflicting ideological constructions inspired by them. One is the dramatic changes in the social structure of Europe's population during the thirty years of unprecedented growth after the Second World War, a transformation particularly important in Europe's Latin countries, where the first phase of the Industrial Revolution had not gone as far as it had, say, in Britain or Germany. Within a generation, a semirural country like France, the proverbial land of small property owners and shopkeepers protected by a high tariff wall, was turned into a country of the vanishing peasant, a land of wage and salary earners having to cope with the full blast of international competition. The evolution toward the pattern inaugurated by Britain and now dominated by the United States, the argument runs, was bound to have political consequences. It announced the "end of ideology," consensus politics, the electoral search for the middle of the road, and so on. According to this version, the great upheaval of the French students *and* workers in May 1968 marks the symbolic end of an era and not the beginning of a new one.[3]

But the other trend is equally undeniable and now holds sway. All of Europe's economic "miracles" are over. Spectacular growth has given way to stagnation and mass unemployment, threatening even that great postwar achievement, the welfare state, which once looked like the pillar of the reformed society. The new conservative leaders, such as Margaret Thatcher, seem to have sensed this sea change faster than did the doctors of the European Left, who continued to prescribe the mild remedies of yesterday to a patient in need of stronger medicine. In this context, the failure of Mitterrand might reflect the inability of the French Socialists to adapt to a new era and thus might contain a very different lesson from the one commonly drawn. It may well spell the doom not of socialism—that is to say, the search for a radically different, alternative society—but of social democracy; in other words, it might mean the end of the belief that such a society can be built gradually, without any qualitative break, within the framework of existing institutions.

Here we are getting close to one of the fundamental problems of our era: Is it at all realistic to talk of an alternative? If it is not, we may have personal solutions but no collective choice. We must simply resign ourselves to a tough time while capitalism resorts to its classic methods for a cure—"restructuring" the economy, getting rid of "excess capacity," destroying value, and throwing millions out of their jobs—and hope that in the meantime, it will discover engines of expansion and long-term growth to replace King

Car and the by now standard consumer durables. Much is at stake, and it is therefore crucial to know whether the Mitterrand experiment failed because it had no chance whatever or because of the way in which it was conceived and carried out. Was the experiment metaphysically doomed, or was it just bungled?

While always keeping in mind that the specific French experience is an illustration of a general dilemma, we shall examine it from two main angles. Domestically, we shall try to determine whether it was possible to change society radically *from above* by simply stepping into the existing Gaullist institutions, or whether the attempt was condemned in advance without persistent pressure *from below?* Internationally, we shall ask ourselves whether the very idea of socialism in a medium-sized state is a serious proposition and, if not, how one embarks on a socialist venture within the broadly liberal framework of Europe's Common Market?

All these questions, unfortunately, are not only relevant, but also urgent. Left-wing governments were thrust into office in France, Greece, and Spain because their conservative predecessors had been palpably unable to cope with the major economic crisis to the satisfaction of the majority of the population. If they, in turn, fail to manage the crisis, and if this occurs all over Europe, the consequences might be much more dangerous. As we shall see, mass unemployment and "Marxists" in office produce quite an explosive mixture. *Pace* Hegel and Marx; history need not repeat itself as a farce. It may be reenacted as a tragedy, although it is fair to add that the number of jobless and their fate are, for the time being, not comparable with what they were in the 1930s.

Here again we have two schools of thought and two possible conclusions. According to the first, experiences of this kind are indispensable in such countries as France and Italy to dispel illusions and teach the workers realistic limits to their resistance; they must grasp what the system will bear and not look beyond. Mitterrand will thus have played the part of a Harold Wilson or a James Callaghan in Britain. Soon the political spectrum throughout Western Europe will be practically limited to a choice stretching from chastened social democracy to a Reaganite Right. It is symptomatic that Mitterrand is being hailed as a "normalizer" and the Democratic party is being offered as a model for the French Left. If there is a future in this version, it is American.

The other school replies that, far from foreshadowing such a spread of conservative moderation, the failure of Mitterrand reveals the bankruptcy of moderate solutions, the vanity of the allegedly easy roads to the New Jerusalem, a myth that must now join in the dustbin the capitalist legends of permanent growth and prosperity for all. Belying the prophecies of the fashionable pundits, the French experience, once it has been absorbed, prepares the ground for an explosive polarization of European politics.

Before we venture answers to these questions, we must look at the facts. To do so, we first glance at the European background against which the French events acquire their historical significance.

I

The Shifting Context

> The time is out of joint; O cursed spite,
> That ever I was born to set it right!
> Shakespeare, *Hamlet*

1

Changing France in an Expanding Europe

Stereotypes survive. The traditional image of a Frenchman wearing a beret and carrying a *baguette* lives on and quite understandably so, since although the beret from the Basque country is by now out of fashion, you can still buy from your Parisian baker bread baked freshly three times a day. The comparatively elaborate cuisine in most French restaurants and the open-air food markets in all districts of Paris, where you can buy, together with meat, fruit, and vegetables, innumerable cheeses and pâtés galore, preserve the impression that there is at least one country in which the capitalist "joy of accumulation" has not yet replaced the feudal "accumulation of joys."

If you dig a little below the surface, however, the old order is crumbling. The two-hour lunch break, which allowed the middle-class father—once the only breadwinner or, if you prefer, cake provider in the family—to see his small children, does not make much sense in the big towns, with their distances and their traffic jams, and it progressively is giving way to a shorter interval or even to a continuous workday. His wife, who now usually works outside the home, has no time to prepare a pot-au-feu or any other dish that takes long to cook and requires a presence in the house; she is likely to buy a steak, since she is among those who can afford it. The shopkeeper in the open-air market is no longer a peasant or even a vegetable grower from the surrounding countryside. More often than not, he is an intermediary who buys fruit and vegetables from a wholesaler. Incidentally, the wholesale market, the "belly of Paris," is not, as it was in Zola's time, in the heart of the capital, a stone's throw from Notre Dame; it is now outside, at Rungis, which trucks can reach by freeway.

There are more obvious signs of the disappearing old order. The local grocer is being squeezed out by the small superette. At both ends of most French towns of medium size are shopping centers that closely resemble their American model. Self-service restaurants have surreptitiously spread throughout the country, and even if Curnonsky, the prince of gourmets, must be turning in his grave, McDonald's and other hamburger establishments are invading the main streets of Paris and of the provincial towns. The resistance may be somewhat stiffer in France than elsewhere in Europe because conditioned reflexes do not disappear at once, and most of the French are still not very distant from their peasant background. The reluctance to accept the new pattern and the almost inexorable advance of that model have a common explanation. The reshaping of France's social landscape is at once fairly recent and fairly thorough. A few figures are needed to convey the unsuspected scale and speed of that upheaval.

On the eve of the Second World War and on its morrow, one in three people who had an occupation in France was working on the land. As late as 1954, agricultural workers still accounted for more than one in four people in the work force. Thirty years later, the ratio had dropped to less than one in ten, farming accounting for about 7 percent of the French labor force. True, in France, like everywhere else, many more people live in the countryside than work on the land, and this rural population, after years of steady decline, has recently shown signs of recovery. It has done so partly because French town dwellers, too, get converted to commuting from beyond suburbia and partly because the once-massive migration from the land was reduced to a trickle in the past decade as a result of the economic crisis. People go to town in search of a job, not, if they can help it, to lengthen the line of the unemployed. Nevertheless, the historical trend is unmistakable. The share of farming in France's labor force is now roughly the same as it was in the United States in the early 1960s and in Britain immediately after the war.

If France is following the trail opened up by Britain, it is doing so in its own fashion. Its peasants were not uprooted by a vast enclosure movement. Indeed, in the nineteenth century and really until the Second World War, they were still being artificially protected by a special tariff wall for political purposes, in the hope that they would act as a conservative backbone within a society otherwise threatened by upheavals. Their subsequent transformation and partial disappearance have also taken specific forms. The concentration in French farming is by now greater than is generally assumed. The relatively large units of over 50 hectares (approximately 125 acres), while representing only 15 percent of all farms, account for 46 percent of the land under cultivation. By the standards of Continental Europe, this is a fairly high proportion. In Britain, however, large farms account for 33 percent of the total and for over 82 percent of the land.[1] There, they are the only farms that matter. Similarly, about one in three British farms employs hired labor, compared with one in ten in France. In fact, farm laborers have vanished from the French countryside even faster than the farmers themselves. Herein lies the peculiarity of this postwar agrarian revolution. Like the British one, it, too, was spurred by the introduction of capitalist methods of production into the countryside, by the supply of fertilizers and of tractors and other equipment, which meant that in France since the war, productivity has been rising faster in agriculture than in industry. But, unlike its British predecessor, the French agricultural revolution has allowed so far for the survival of the, admittedly modernized, family farm.

Between the gradually disappearing small peasant, often an old man allowed to survive at the subsistence level,[2] and the estates that make a handsome profit are the middling farmers, who own some 20 to 50 hectares (roughly 50 to 125 acres). They are mostly former small farmers who borrowed heavily to buy additional land and, more generally, to modernize in order to survive. Collectively, they own just under 33 percent of the farms and over 38 percent of the land. There is no doubt that in France, they still matter; the only question that arises is over their social function. Some commentators insist on their dependence at both ends. With their land mortgaged and their equipment bought on the installment plan, the pattern of their output is totally dictated by the wholesalers of the industrialized food complex. They can, in a way, be likened to sweated domestic labor, which owns and maintains the tools of its own exploitation. But they are the nominal owners of land and share with their ancestors the "talisman of property." This may explain the almost schizophrenic dichotomy of their behavior. When they think that the commodity prices set in Brussels by the commissioners of the European Economic Community (EEC) will not allow them to make ends meet, they drive their tractors into town, pour potatoes or artichokes into the streets, let pigs loose, and fight with the police who are protecting official buildings. In between such explosive bouts, they are the staunch upholders of the conservative order. At least they were until now.

Whatever the long-term future of this new figure in the French countryside, the middling modernized farmer struggling for survival, the drastic reduction in the total number of farmers—from nearly 4 million after the Second World War to barely 1.5 million in 1982—has greatly contributed to the other, more general feature of postwar French society, the steady decline in the number of *indépendants*—whether small landowners, artisans, or shopkeepers—within the total working population. In 1954, the growing army of wage and salary earners still absorbed only 65 percent of the labor force. Thirty years later, its ranks had grown to 82 percent of the total. Admittedly, the process has been slowed down by the economic crisis, yet rising above temporary vagaries, one must conclude that the French are catching up with the Americans and the British (the proportion is 90 percent in the United States and 95 percent in the United Kingdom).

And they have been catching up at a time when their army of labor began growing once again after decades of stagnation. In 1962, the French work force, at just under 20 million, was the same size as it had been at the beginning of the twentieth century. Under the combined influence of the postwar baby boomers reaching the labor market, foreign workers flooding the country, and women going out to work, the work force started to grow at a steadily faster rate. For a long time, this increment, as well as the surplus provided by migration from the countryside, were easily absorbed by the other sectors of the economy. In the late 1960s, construction ceased to expand. By the mid-1970s, it was the turn of industry to contract, the services alone dragging total employment slowly upward. In the 1980s, the fewer jobs still provided by the services were no longer enough to compensate for the losses occurring in all the other sectors. Although immigrant workers had long ceased to come to France, the continually growing labor force merely fed the by-now vast army of the unemployed.

Seen in perspective, over a period of nearly thirty years, the changes are more revealing. Between 1954 and 1982, employment in farming went down by 66 percent; in industry, it first climbed and then dropped, rising by only 10 percent; in the services, it grew all the time, altogether by over 80 percent. The same figures may be presented even more dramatically: In 1954, out of 100 people working in France, 27 did so in agriculture; 35, in industry (including building); and 38, in the services. In 1982, there were about 8 in farming, 34 in industry, and 58 in services.

In this metamorphosis, France is far from exceptional. It fits in with the prevailing Western trend. The snag is that the services—or the "tertiary sector," as the economists like to say—by contrast with the primary (agriculture and mining) and the secondary (industry) sectors, are a hodgepodge, combining millionaire managers with their office cleaners and, on their own, telling us little about the real movements within society. A closer scrutiny is required to determine where jobs have been created on a massive scale. Undeniably, in the social services, primarily in education and health. Since in France, teachers at all levels and the medical staffs in hospitals are mostly public servants, as are the post-office employees, who benefited from the development in communications, the public sector, the central and local authorities combined, has become the main employer, accounting for roughly 30 percent of all wage and salary earners—and this without taking into account the industries nationalized by the Socialists after 1981.[3]

Other expanding sources of employment are less clearly defined and circumscribed. They include management and supervisory functions, as well as jobs for technicians and researchers. They cover all kinds of clerical and commercial occupations in banking or insurance and all sorts of services rendered to firms from outside in an ever increasingly complex division of labor. It is just not right to interpret the faster expansion of the tertiary sector as the replacement of blue-collar by white-collar workers, simply because there are many blue-collar workers in the service industries. Altogether, to grasp the changes in the class structure of society, it might be better to comment briefly on the data of what the French call "socioprofessional categories," which are summarized in Table 1.1.

Among the farmers, in 1982, only 335,000 were considered by the statisticians to be fairly big ones. In that year, there were 131,000 nonagricultural employers with more than 10 wage earners; out of these, 95,000 had fewer than 50 employees and 30,000 had fewer than 500 employees, leaving about 6,000 owners of very big firms. Among the small *indépendants,* the ranks of the shopkeepers were more rapidly reduced (by over 30 percent in 30 years) than those of the craftsmen (by 25 percent over the same period); artisans even recovered slightly in recent years, since starting a business of one's own began to be considered as a palliative against unemployment, particularly in the building trades, which account for over a third of all independent artisans.

The next big category, and a fast rising one at that, centers around that awkward and all-embracing French term *cadres.* In the late eighteenth century, *cadres* was the name

Table 1.1 Changes in the French Labor Force

	'000		Share (%)		1982
	1954	1982	1954	1982	1954 = 100
Farmers	3,966	1,448	20.7	6.2	36.5
Farm workers	1,161	304	6.0	1.3	26.2
Industrialists and traders	273	281	1.4	1.2	102.9
Craftsmen and shopkeepers	2,029	1,456	10.6	6.2	71.8
Higher management and liberal professions	554	1,810	2.2	7.7	326.9
Lower management, teachers, technicians, and medical staff	1,113	3,254	5.8	13.8	292.5
Shop assistants	441	931	2.3	4.0	211.3
Office workers	1,628	3,745	8.5	15.9	230.1
Workers	6,490	8,266	33.8	35.1	127.4
Personnel and domestic service	1,018	1,531	5.3	6.5	150.5
Total (including artists, clergy, army, and police)	19,185	23,525	100.0	100.0	122.6

Source: Economie et statistique, nos. 171–72 (Paris: Institut National de la Statistique et des Etudes Economiques, November–December 1984), pp. 156–57.

given to the officers, commissioned or noncommissioned, in charge of troops, and today, too, the term refers broadly to the higher and lower officers who command the army of labor, except that because of their social proximity to these managers, doctors and professors have been assimilated into this group.[4] One may, therefore, define the *cadres supérieurs* as the professional upper middle classes and observe that they have grown faster than any broad section of French society, more than tripling their ranks within a generation. (This may be partly due to factory or shop owners registering as *cadres* for tax purposes.) The *cadres* in the narrower sense of the term, the higher managerial staff in the private sector and the less numerous top civil servants, did not expand quite as rapidly. The number of engineers more than quadrupled, admittedly starting from the very low level of 75,000 in 1954. Among the groups included by analogy, the liberal professions—mainly doctors and lawyers in private practice—did not even double their numbers in the 30 years, whereas, at the other extreme, scientific researchers, university professors, and high-school teachers (who in France are called *professeurs* and have a much higher status than in the English-speaking world) beat the record with a sixfold increase, reflecting the educational boom of the 1960s and early 1970s. It is among them that can be seen the signs of the social advance of women, a subject to which we shall return in a moment.

The middle-class professionals, or NCOs, grew slightly less rapidly than the higher ranks, their numbers increasing threefold over the period. Among them, the development was uneven. The middle-rank administrative staff in business management and marketing (including traveling salespeople) as well as the intermediate ranks in the civil service only doubled. The number of technicians was multiplied by nearly five, even if their expansion has now slackened, except for those working in electronics and computers. The big boom among primary-school teachers is now over, too, but the junior staff in the health and social services—qualified nurses, social assistants, and therapists—continues its rapid advance, and because women are overwhelmingly predominant in this group as well as in elementary-school teaching, they now represent roughly half of this professional middle, or, rather, lower middle, class.

Women actually account for two thirds of the next category, the employees,[5] combining the lower rungs of office workers with shop assistants. Together, they have grown steadily rather than spectacularly. Still, there are twice as many today as there were a generation ago. In the French definition, the employees naturally include secretaries, bank and office clerks, junior accountants, telephone operators, draftsmen, and so on, but also holders of less clerical jobs in the post office or in the hospitals (medical help). They are a vast congregation, predominantly young and female, adapting themselves all the time, even if the name remains the same, to the shifting requirements of a changing economy. The rise in the number of shop assistants, contrasting with the decline in that of shopkeepers, merely reflects the concentration in French trade, a process that would have been faster still but for some legal constraints that slowed down the spread of supermarkets. Altogether, the employees are numerically second to none except the industrial workers. They have also much in common with the workers in terms of status or of alienation in their job. A woman operating a cash register in a supermarket could very well play the main part in an up-to-date remake of Charlie Chaplin's *Modern Times*.

The share of the workers in the labor force was slightly higher at the end than at the beginning of the period. Actually, it went up until the late 1960s and has declined since the mid-1970s. This may surprise the reader who has recently heard so many farewells to and funeral orations over the coffin of the industrial proletariat. Numerically and socially, the working class remains crucial in most countries of Western Europe, even if the relative weight of blue-collar as opposed to white-collar workers has undoubtedly declined; the relationship between the two should really be reexamined in the context of current relations of production and exploitation.

Unskilled jobs have been, so far, the main victims of the

new economic crisis. Does this mean that the advent of the robot heralds the death of the laborer? And how provisional or permanent is even this relative decline of the industrial proletariat? Its inner prospects and changes will figure prominently in this story. Here, it is enough to give a final word of explanation about Table 1.1. The increase by half of those employed in personal services should not be interpreted as the development in France of domestics. Their numbers have in fact dwindled. This has been more than compensated by the spread of medical assistants, mother helps, and so on, as well as by the net increase in the number of waiters in restaurants and general staff in hotels. (Even in this "personal" domain, society tends to move from the feudal to the capitalistic.)

The expansion of white-collar jobs has been coupled with the mass entry of women into salaried employment. This second wave of women going to work is a relatively new phenomenon. There were about 7 million working women in France at the beginning of the century, accounting for nearly 36 percent of the labor force. It took almost seventy years for this level to be reached again in absolute terms and a few years longer in proportional terms. Then, between 1968 and 1982, some 2.5 million women joined the labor force, bringing their share in the total to more than four in ten. Statistical comparisons concerning women, however, are awkward. At the beginning of the twentieth century, most women were employed in agriculture. Whether on the farm or in the shop, they tended to work in the family firm. Complications arise from the fact that women, as a rule, hold two jobs. The oft-quoted extreme example of statistical aberration is that of an aged widower marrying his children's governess: Although her job remains the same, she suddenly vanishes from labor statistics. The great difference between the beginning and the second half of this century is that in this second wave, most women went out to work for a wage.

Altogether, the share of working men and women in the total population has been going down as a result of longer education and earlier retirement. For women, this trend was reversed in the mid-1960s as more among them began to go back to their jobs soon after childbearing. Taking women between the ages of twenty-five and fifty-four, in 1968 less than 50 percent of them worked; fifteen years later, nearly 66 percent did so. France again was no exception. The proportion of women of working age actually in the labor force (known as the rate of participation) reached in France 53 percent in 1982. This was slightly higher than in Germany and much higher than in Italy, but lower than in Britain and much lower than in Scandinavian countries, where many more women had an opportunity to do part-time work. France also lagged substantially behind the United States (a rate of 61 percent), where part-time work was not much more common than in France.[6]

The link between the expansion of services and the employment of women is obvious, since by 1982, three out of four women earning a wage in France were in the tertiary sector, compared with only two out of four men. The patterns of employment are very different for men and for women. In 1982, out of 100 male wage earners, more than 50 were still blue-collar workers; out of 100 women, only 20 had blue-collar jobs, while 37 were employees.[7] Women are overwhelmingly dominant in domestic and personal services. They outnumber men considerably among employees, whether as shop assistants—particularly in food, clothing, and luxury goods—or as typists, secretaries, and junior clerical staff in both the private and the public sectors. Although very rare among foremen and technicians, they are more numerous than men in primary-education, health, and lower managerial jobs, thus preserving their general average—four out of ten—in the intermediate professions or lower-middle-class jobs. But they account for only about 25 percent of the total both among workers and within higher management.

On closer scrutiny, the position of women is less rosy than these figures might suggest. They are, for instance, four times more numerous among the unskilled workers than among the skilled ones. Their gains in upper-class professions are due almost entirely to their entry into teaching and medicine. There are more women than men among high-school *professeurs* and almost as many among salaried doctors. The proportion drops to less than 30 percent among university teachers and scientific researchers. On the whole, women are doing better, or less badly, in professions for which diplomas are required and examinations have to be taken. Their financial handicaps are somewhat lower in the public service, where the average wage earned by women lags around 20 percent behind the wage earned by men, compared with 42 percent in the private sector. Although barriers are breaking down, with women gaining ground among dentists, medical specialists, and lawyers, the Bastilles still to be stormed are many. Nevertheless, undeterred by the economic crisis, the tide of the working women looks inexorable.

The same cannot be said of the other contingent that played a significant part in Europe's postwar "miracle"—the foreign laborers, called by the Germans, with a strange sense of humor, the "guest workers." Although obviously unwelcome "guests" since the beginning of the economic crisis, foreign workers have come to stay. They have become a structural part of the European economy. In the present period of permanent unemployment, however, their numbers are likely to decline rather than to grow.

Immigrant workers are Europe's white niggers. Coming first from the poorer countries of Europe to the richer ones and then from beyond Europe's frontiers, they were brought in during the boom years to supplement peasants migrating to towns and women joining the army of labor. Uprooted at home and driven abroad in search of work, they were ready to take on jobs considered by natives as badly paid, dirty, monotonous, dangerous, or unhealthy. They cleaned the houses and swept the streets, and they replaced women on night shifts in the textile industry or men on assembly lines in the car plants. In the 1960s, when foreign workers reached one third of the labor force in Switzerland, Europe looked as though it were on the road back to the model of ancient Greece, with leisurely citizens and productive slaves. The crisis partly reversed this trend. Nevertheless, something like one tenth of the Western European working class is disfranchised.[8]

France, for all the publicity surrounding Jean-Marie Le Pen and his National Front, is no champion of immigration. With foreigners accounting for 6.8 percent of the total population in the last census and immigrant workers for roughly the same proportion of the labor force, France was close to the Western European average, halfway between Switzerland, where the share of foreign laborers in the working population has now dropped to around 15 percent, and Italy, where, if one may say so, immigrant workers are really natives, southern Italians gone to the industrial north, to Turin and Milan, in search of work. Nor is the present proportion a French record. In a country where the population was stagnant almost from the time of Napoleon I, immigration has a long history and the ratio of foreigners to natives had reached the same level in 1931.[9] Finally, the recent outcry has nothing to do with a sudden influx. Drastically reduced after the Second World War, the foreign population climbed rapidly during the years of economic expansion, rising from 4.1 percent in 1954 to 6.5 percent in 1975. Since then, with the frontiers virtually closed to immigrant workers from outside the European Economic Community, the share has risen only fractionally because of family reunions. The reason for the revival of xenophobia is the rise of unemployment and the racist pretext—the shift among immigrants from Italians and Spaniards to Arabs from North Africa.

What is naturally forgotten in this new mood is the contribution these "guest workers" made not just to Europe's "miracles," but also to the social rise of their hosts. The natives did climb on their backs. The white-collar "revolution," that of "rising expectations"—or, to put it less eloquently, the steady advance of the mass of Western Europeans up the social ladder—was rendered possible partly by the arrival of these outsiders resigned to occupy the lower rungs of that ladder. And for once, the advance was genuine. The quantitative changes in the pattern of consumption were so sweeping and so rapid as to have a qualitative effect on the way of life. The American dream, first glimpsed as the nylon stocking in the GI's knapsack and subsequently symbolized by the idea of an automobile standing in front of a workingman's house, became a European reality within a generation, within the working life span of that man.

Measured in material terms and official figures, real wages roughly tripled in France within a third of a century. The fast growth of consumer durables, taken in the postwar era as *the* yardstick of living standards, confirms the verdict.

As late as 1960, a car was still a privilege, since less than one household in three possessed one; the refrigerator or the washing machine was a luxury, owned by one in four; and the television set and the telephone were the real exceptions, absent in nine out of ten houses. A quarter of a century later, if the size of the goods may not be quite the same, the statistics are almost American. The washing machine and the refrigerator, the telephone and the television are found in nine out of ten French households, and the ratio would be the same for the car if you ruled out households of the very old and the single.[10]

The spread in the ownership of durable consumer goods should not be identified with a leveling of living standards. Durables are not all. One car is not equal to another, while the cost of running a car—estimated at 13 percent of a French family's budget—is not the same burden for the boss of a company and an unskilled laborer. Actually, during the first half of the period under consideration, the gap between the average wage of a worker and the average salary of a manager did widen. The year 1968, when the strikers won a big raise in the national minimum wage, marks a reversal of this trend. Subsequently, the differential narrowed. Looking at total net incomes after taxes, the economic crisis brought, in a sense, a leveling because of swollen social transfers, due to unemployment and early pensions, as well as surtaxes to cover the expenditure; this higher cost has actually raised the question of the future of the welfare state. But lasting unemployment also widened the gap, thrusting to the bottom a reinvented category—the new poor.

In a more general way, dazzling from a distance, the American way of life did not look quite as impressive at close quarters and did not prevent discontent or social upheaval. We shall see that the improvement in living standards was not a steady, regular movement but one involving a phase of breathtaking acceleration followed by one of painful stagnation. Despite all these reservations, the thirty years after the last war were ones of unprecedented growth, of tripling living standards within a generation, and of a radical transformation of the European way of life.

Let us stop for a moment to summarize the main features of Europe's social upheaval, taking France as a striking example. Its rural landscape, we saw, was reshaped beyond recognition by the virtual disappearance of the peasant. The reduction in the number of the small property holders in the towns—the shopkeepers and artisans—was incomparably slower. During the boom years, the mass migration from the countryside was not sufficient to satisfy the demand for manpower. Immigrant laborers, coming from increasingly distant lands, and women joining the labor market in growing numbers made up the difference, the employment of women going hand in hand with the development of the so-called tertiary sector. Over the period as a whole, the metamorphosis was not the one fashionably painted, of an industrial society moving to the postindustrial age. Instead, it was a major shift of manpower from agriculture into the services in the broad sense of the term, with industry, although far from stagnant, employing about as many people at the end as at the beginning.

All these shifts and permutations had consequences that are still spreading through the body politic. The French Right, for instance, as the upholder of private property, has had to adapt its policy and discourse to the evolution of large sections of its electorate from small landowners to salary earners. The Left has shown less ability to take into account the inner transformation of the working class. The labor unions, in particular, have proved so far incapable of coping with the shrinking of their traditional fiefs or the inclusion of women and immigrants into their ranks. Yet there is another political factor that cannot be left out—the role of the state, which evolved in quite contrasting fashion. On the one hand, its powers widened tremendously. The so-called Keynesian management of the economy did not mean just the use of the budget as a policy instrument. The state became directly involved in the fields of education, health, and insurance. It acted as a sophisticated supervisor of the reproduction of labor. On the other hand, the capacity of the nation-state to control its own economy was being undermined by the continued progress of the international division of labor, and this tendency was permanent because the postwar expansion in Europe was coupled with—nay, was aided and abetted by—the opening of frontiers.

In the years when production was booming, foreign trade was growing even faster. Thus between the birth of the Fifth Republic in 1958 and its first political crisis ten years later, the gross national product of France rose by three quarters and its foreign trade doubled. Taking the ratio of exports to gross domestic product as a rough measure of the importance of foreign trade in a country's economy, it rose in France from over 13 percent in 1955 to around 15 percent in 1960, over 20 percent in the mid-1970s, and close to 25 percent ten years later.[11] This is not much compared with the share in small trading countries, such as Belgium, and it is slightly less than the proportion in Britain or Germany, but three times the level of the United States in the 1980s.

The opening of the frontiers was even more dramatic than these figures suggest, since much of the French trade used to be carried within the sheltered confines of the colonial empire. In 1958, the inaugural year of the Common Market, France's five new European partners accounted for

22 percent of its foreign trade, and the so-called franc area for over 33 percent. This was still the time of the Algerian war, and it may be argued that to bring that war to an end and extricate France from its colonial venture was by then in the interests of the dynamic sections of French big business.[12] Ten years later, the positions were more than reversed. The share of the former overseas possessions had gone down to 13 percent, and that of the European Economic Community up to 43 percent.

For its six original members, the EEC offered the prospect of a big market, the scope for a more efficient division of labor and for economies of scale. The idea of some of the federalist inspirers of the Rome Treaty, who, keeping the reunification of Germany in mind, thought that the *Zollverein,* the customs union, might lead fairly soon to a Western European state, did not come true. As a customs union, however, as a commercial common market, the EEC proved a success. In a relatively short time, a country like France had to abolish trade barriers with its partners, but also to lower duties for the rest of the world in order to fit into the outer tariff of the EEC. While all this stimulated production and trade, the liberal framework of the Common Market was bound to raise a series of problems for any member country determined to carry out really radical reforms.

The story of Europe's postwar social upheaval, as told briefly here, is clearly too contradictory, too full of ifs and buts. Inevitably so. To convey the extent to which European society has been torn apart and reshaped since the Second World War, it was necessary to treat the forty years as a whole, whereas they covered two phases that were very different from each other in length and nature: three decades of exceptional prosperity, followed by one of no-longer-customary depression. Journalists in search of sensational headlines wrote first about miracles. It began with the "German miracle" in the early 1950s, attributed to the inflow of labor from the eastern part of the divided country and to the hard work characteristic of that nation. Then came the "French miracle," perceived when the temporary stabilization of the franc revealed the dynamism of the French economy. When the same miracle was discovered in Milan, it became obvious that something deeper was at stake than a series of national successes. A combination of factors—the spread throughout Europe's industries of a modernized version of Taylorism; a greater intervention of the state to spur production, boost consumption, and regulate social life; and the opening of frontiers—drove the European economy forward at an unusual speed. Even Britain, the odd nation out, advancing at half the pace of other countries, was still growing at an annual average of 2.5 percent—that is to say, much faster than during its heyday in the nineteenth century. And this exceptional expansion was not limited to output. Partly under pressure from the trade unions, partly to induce the labor movement to collaborate more closely with the establishment, yet also to ensure a market for its goods, the economic system encouraged the development of mass consumption. The rise of living standards, too, was unprecedented, leading to a radical shift in the very mode of life.

Presented only through impressive statistics, this period may seem in retrospect more attractive than it actually was. The American dream was not exactly what it had been made out to be. The uprooted peasants, the squeezed shopkeepers, the migrants doing dirty jobs, the women doing two jobs at a time, the workers having to keep up with the infernal rhythm of their machines—all these people were not in seventh heaven in the overcrowded towns bursting at their seams. The consumer was no king in the so-called consumer society, merely an addict and a victim. The iconoclastic students were the first, in the 1960s, to question the gospel of growth: Growth for whom, in whose interest, and for what purpose? Young workers responded, stressing the other side of the coin: the mad pace of the assembly lines and the alienation of fragmented, parceled-up labor. The magic was broken. The Western world was suddenly full of recrimination about the loneliness, the meaningless nature, and the absurdity of modern life. The French uprising of 1968 and the Italian social upheaval in the following year revealed the depth of discontent beneath the glittering surface of European society.

Nothing fails like failure. A rebellion that does not shatter the structure of existing society is inevitably stifled and integrated into that society. Radical breaks, although crucial in history, are rare. They are the exception, the brief interval. During the long periods in between, old customs, the dead weight of the past and of the established order, dominate conservative societies. Still, it was not easy to stifle at once, or bottle up, the rebellious spirit of the 1960s and destroy the glimmer of hope aroused by the prospect of a different future. In Western Europe, in particular, the political parties of the Left found it difficult to convince their supporters that there are no revolutionary shortcuts to a radically different society and that life can be altered gradually, progressively, within existing institutions. The statistics mentioned earlier and the reality behind them played a major function in this process. The workers, after all, had cars and refrigerators and not only their chains to lose.

History has its ironies and paradoxes. The irony is that by the mid-1970s, when the gospel of growth was finally restored as the ruling religion, it had become obsolete. The "miracle" was over. The paradox is that the resulting crisis of capitalism—a major one, not necessarily fatal, yet requiring a wholesale revision of the mechanism of accumulation—far from weakening the natural upholders of the system, the conservatives, the Right, had bewildered its official opponents, the parties of the Left. Maybe the paradox is only apparent, since new converts are usually less skeptical than the old faithful.

2

The Left Bewitched and Bewildered

The end of the era of seemingly permanent growth took almost everybody by surprise. The conservatives in the West reacted more swiftly than the Left, dropping at once any pretenses about a search for consensus. They resurrected class conflict at its most naked and revived the open glorification of the capitalist jungle, successfully presenting this worn-out myth of yesteryear as the quintessence of modernity. The Western Left, in contrast, trying desperately to stick to the middle of the road, as though nothing had happened, seemed to be hopelessly lost. In fairness, it must be said that the European Left had some justification for its dizziness. Within a fairly short spell, it had lost a model, a comfortable pattern of collaboration, and now, through the crisis, big chunks of its traditional support.

Thirty years, depending on how you look at them, are a historical instant or the life of a generation. Nikita Khrushchev's famous "secret speech," his indictment of Stalin, was delivered in 1956, which is not so long ago, yet many people can neither remember nor imagine how monolithic the international Communist movement was at the time. Big parties, with deep, genuine roots in their national soil, acted like puppets dancing to the latest Kremlin tune and talked collectively like a ventriloquist's dummy. The discipline, while imposed from above, was easily accepted: Millions of Communists were convinced that theirs were mere regiments in a vast revolutionary army, whose main battle front was in the Soviet Union, where an alternative society was being built despite enemy blows. Whatever the odds, the setbacks, or the vicissitudes, the Russian comrades were forging the future.

True, while Communists formed big battalions in France and Italy, they were reduced to tiny detachments, some might say marginal sections, in Britain and Germany (where the Communist party was, incidentally, banned). Whether the Soviet experience taken as a whole, from its revolutionary beginning to its post-Stalinist sclerosis, has helped or hindered the spread of socialism is highly debatable. The pioneering feat of the Bolsheviks—the proof that the workers could seize power and hold it, the abolition of private property, or the planned shortcut to industrialization—has inspired millions of people throughout the world. The bloody collectivization, the association of socialism with Stalin's crimes, with the "concentrationary" universe and of the soviets with the absence of workers' democracy have turned many others against the very idea of radical change. The balancing act is awkward, and since there is no possible experimental proof which weighs more on the scales—the achievement, or the cost—historians will ponder it for generations, as they will argue whether Marxism, iconoclastic in its very essence, could be confused with the Byzantine cult proclaimed in Moscow, allegedly in its name, or whether this cult, designed for backward *muzhiks,* was bound to disappear, Khrushchev or no Khrushchev, once the system itself had uprooted the *muzhik.*

All these are fascinating and controversial problems that, fortunately, need not be tackled, let alone solved, here simply because the only point we are trying to stress is that the breakup of the once-monolithic international Communist movement is a fairly recent phenomenon. February 25, 1956, is a crucial date not because of the historical or theoretical content of Khrushchev's oration, which lasted well into the night. The facts he quoted about Stalin's sins brought almost nothing new to Western experts and contained little likely to sway the general public interested in Soviet history. (The "discovery" of the gulag twenty years later through the works of Aleksandr Solzhenitsyn, we shall see, had less to do with the sudden availability of information than with the change in the political climate.) As a piece of Marxist analysis Khrushchev's explanation of the ills of a system and a society as a consequence of the psychopathology of one evil man or, rather, a good ruler gone wrong is not even worth the paper on which it was written. It is a major historical document for quite different reasons.

For millions of followers, attacks against Stalin's Russia were simply anti-Soviet slanders, and they had grounds for doubting the objectivity of capitalist propaganda. Now, however, the accusations were being confirmed by an unimpeachable source. The cult was being denounced by its official keeper, Stalin's own successor as the leader of the Soviet Communist party. Thus supporters of Soviet Communism had lived a lie and deified a monster. The revelation had a traumatic effect on the faithful. Khrushchev's words shook not just Russia, but also the whole Communist world.

The reverberations were felt first within the Soviet bloc, the words in Moscow producing an immediate echo throughout Eastern Europe. In Poland in that same year, during the so-called Spring in October, the Russians reluctantly accepted the return to power of Wladyslaw Gomulka, showing thereby that they were resigned to a reform of their empire. The Polish events, however, had at once a counterpoint in Budapest, the Soviet invasion of Hungary spelling in blood the limits of tolerable change. Nevertheless, even after this second shock, Western Communists found reasons not to despair. Do the rulers often reform a system on which their power rests by extending the frontiers of freedom, as Khrushchev and his comrades had at least begun to do? The worst features of Stalinism, they argued, were the result of Russia's backwardness; yet Stalinism was at the same time the instrument of Russia's industrial revolution. In the land of the *Sputnik*—Yuri Gagarin, the first cosmonaut, took off in April 1961—inexorable economic progress was bound to pave the way for the spread of democracy. But after the short spell of thaw, winter returned, milder than before, although seemingly interminable. It took Leonid Brezhnev's freeze—eighteen years of stability for the ruling bureaucracy, purchased at the price of total absence of political and economic reform—to shatter the hopes, bred by a mechanical Marxism, of inevitable political change. In the meantime, the stifling of Czech freedom, the entry of Soviet tanks to nip in the bud the Prague spring of 1968, killed the remaining illusions about a smooth, orderly transition, guided by the leadership, from Stalinist empire to socialist democracy.

The reaction of Western European Communists to the initial shock was typical. The French tried to ignore it by putting their heads into the sand: For years, they referred to the famous indictment as "the report attributed to comrade Khrushchev." The Italians took advantage of the new situation to begin distancing themselves from Moscow. It was their leader, Palmiro Togliatti, who, if he did not invent it, popularized at the time the concept of polycentrism.

The idea was clever and convenient. If Moscow were no longer the Communist Rome, the capital and headquarters of the international movement, it was not so urgent for Communist outsiders to study the Russian heritage, probe the roots of Stalinism, and, in the process, ask current Soviet rulers and Stalin's past lieutenants what their roles had been in the performance of their master. Polycentrism enabled Russia to be treated as a special case and Stalinism as a product of peculiar circumstances. What the Soviet leaders were being asked, in fact, was to stop making of necessity, however horrible, a virtue and of their policy, a categorical imperative for the movement as a whole.

In thus outlining a break with the Soviet model, polycentrism foreshadowed another ism, which was to have its vogue two decades later—Eurocommunism. A seven-day wonder, this nickname hit the headlines of the world press when the leaders of the three main parties of this movement—Enrico Berlinguer of Italy, Santiago Carrillo of Spain, and Georges Marchais of France—met in Madrid in March 1977. Geographically, the concept was rather ill-defined. If Euro- stood clearly for Western Europe, the Japanese Communist party was also described as Eurocommunist at the time. Very vaguely, the term referred to Communist parties that were relatively independent from Moscow and were considered to be liberal, whatever that term really meant, both in their strategy and in running their own affairs.

Keen to lose the Stalinist label and not to be branded as approving, say, the treatment of dissidents in psychiatric wards, Western Communist parties were busy rewriting programs and brushing up their vocabulary. They became as eclectic as they had been narrowly dogmatic in the past. Thus Marchais explained to his French followers the dropping of the concept of the dictatorship of the proletariat on the grounds that the very word *dictatorship* had bad connotations and reminded ordinary people of Hitler and Mussolini (although, strangely, not of Stalin). He did not tell the members that for decades their party had approved the term, while in the Soviet Union it had meant something quite different—the dictatorship *over* the workers. Neither did he explain that in the original Communist conception, dictatorship of the proletariat was a vital link covering a whole period of transition from a capitalist to a classless society. He just dismissed it because it did not sound good.

Cutting the umbilical cord was one indispensable move. Changing the nature of the party was the next. Western Communist parties did not inherit the original Bolshevik model but its Stalinist version, in which "democratic centralism" means that all power flows from the top. In circumstances very different from czarist illegality, with purposes other than the seizure of the Winter Palace, the organization of non-Soviet parties had to be different. The enthusiasm to criticize domestic sins and to loosen the hold of the leadership was not so great as that to put the blame on the Russians. To take the Spanish example for a change, Carrillo, a belated but stern critic of Stalinism abroad, is described as quite a practitioner of such habits in his own party. Indeed, according to Spanish specialists,[1] the absence of genuine debate and democracy in its own ranks contributed greatly to the sharp fall of the Spanish Communist party from the principal force in the resistance to Francisco Franco on the eve of his death to quite a marginal influence in Spanish politics ten years later.

Still, for Eurocommunist parties the key to the future lay elsewhere. The break with the Soviet model, however vital and indispensable, was only the beginning. Nature abhorring a vacuum, something had to be put in its place. Rejecting the Soviet system of rule from above—whether by a dictator, a clique, or a central committee—did the parties seek new forms of democracy from below, giving the people for the first time an opportunity to shape their own destiny, or were they content with parliamentary democracy as applied in the advanced capitalist countries? Dismissing the Soviet version of overcentralized planning as bureaucratic and inefficient, did they invent forms of workers' councils and methods for reconciling planning with decentralization, or were they resigned just to limit the vagaries of the market? Critical of growth for profit or for its own sake, did they propose new ways of producing for genuine needs and for inserting humans into the environment? After all, at least in principle and on paper, Communist parties are heirs to the old socialist ideal of a world run by equal and associate producers, a classless society in which the frontiers between town and country, physical and mental work, labor and leisure have disappeared, in which the hierarchical division of labor has vanished and the state has withered away. Had they abandoned the ends as utopian on discovering the horror of the Stalinist means and stopped talking about them because Soviet leaders still pay lip service to these goals on rare festive occasions?

The honest answer is, to use an understatement, that such preoccupations did not figure prominently in the Eurocommunist debate. The explanation is to be found in the general political climate, affected by changing winds from both East and West. For years, Western Communists had looked at the Soviet Union through rose-colored glasses, perceiving only achievements: the staggering rate of industrialization, the spread of mass education and culture, the careers open to yesterday's downtrodden. Their eyes fixed on the Magnitogorsk complex or on cosmonauts in space, they swallowed the ghastly trials in Moscow and turned a blind eye to mass repression. Now they were seeing and accepting the image of Russia normally available in the Western world, which is focused on the seamy side: the long lines, the overcrowding, the shoddy goods, the absence of choice for the consumer, the political impotence of the proletariat in the alleged land of the workers. And to make matters worse, the Brezhnev years witnessed a slowdown in production and a widening of the technological gap between East and West.

While the picture of Russia was thus getting darker, that of the West was becoming brighter and, in many ways, idealized. The comparison was being drawn at a time when advanced capitalism was living through the closing stages of a long cycle of unprecedented prosperity. Was it wise to talk of capitalism as being at the end of its tether when it seemed to have just discovered the secret of eternal growth? Was it reasonable to refer to "absolute pauperization" when Western workers were lovingly shining their cars and installing washing machines in their houses? Admittedly, the euphoria was ephemeral and even then, motives for discontent in the West were deep and numerous. The fact remains that when the Soviet model was being shattered, its capitalist counterpart glittered and the Eurocommunist debate was inevitably influenced by its environment and the prevailing mood.

Naturally, one should not oversimplify. Even in Western Europe, not all the parties turned Eurocommunist. The Portuguese Communist party, headed by Alvaro Cunhal, and the bigger of the two Greek ones, the KKE-ex, the Communist party of Greece, known as "external," went on their way as though Stalin had not existed and capitalism had not changed its spots. The party of Marchais, which we shall meet time and again in this story, managed to get the worst of both worlds: Too distant from Moscow for some of its old-time supporters, it was sufficiently close to the Soviet position, on, say, Afghanistan or later Poland to be found guilty by association by the bulk of the French electorate. Finally, the standard bearer of Eurocommunism proved to be the Italian Communist party. Although also without a clear perspective, it has weathered the storm so far better than its partners: Year in and year out, it musters about 30 percent of the vote in the national polls and remains by far the dominant force on the Italian Left. But it may well be described as having gone—first under the leadership of Berlinguer and, since 1985, under that of Alessandro Natta—well on the road toward social democracy.

There are words that, unless clearly defined, add confusion rather than precision to an argument. *Social democracy,* whether used with contempt or with admiration, is such a term and, therefore, requires a brief introduction. Before the First World War, most Socialist parties belonged to the Second International and were called Social Democratic, including Russia's nonidentical twins, the Bolsheviks and the Mensheviks. The pride and model of the international organization was the German Social Democratic party (SPD), rapidly rising in strength, membership, and parliamentary representation. Like all the other parties of the International, it was committed to the abolition of capitalism and of the "exploitation of man by man," to the establishment of a classless, egalitarian society. The famous controversy within the party between "reformers" like Eduard Bernstein and revolutionaries like Rosa Luxemburg was ostensibly over the means, not over the goals (could the ends be achieved peacefully, gradually, through parliamentary procedure?). Strange though it may seem in retrospect, Rosa the revolutionary was then the champion of orthodoxy, and Karl Kautsky, the keeper of the line, was her ally. However sanguine the party's theory was alleged to be, it was possible to be skeptical about the radicalism of its praxis, although the doubts were not openly confirmed until 1914.

The outbreak of the First World War was a traumatic event for international socialism. In principle, socialist workers were supposed to refuse fighting against their foreign brothers and to turn their weapons against their native exploiters. In fact, most of the leaders of Social Democracy joined the patriotic stampede in August 1914, climbed on the military bandwagon, and became the most zealous supporters of the war. A minority stuck to its principles, paid a price for it, and cursed its former comrades as traitors. Passions rose, and the gap widened as the war proceeded with its terrible toll. The Bolsheviks were, naturally, in the antiwar faction, and the Russian Revolution crystalized the divisions. The creation by the Russians of the Third International, the Comintern, in 1921, fixed them for long, if not forever. To say that at the time the split between the Second and the Third Internationals was one between reformers and revolutionaries is literally inaccurate and broadly true. (By the 1930s, the situation was more complex. The Comintern was by then the not necessarily revolutionary instrument of established Soviet power and the Socialist International a rather loose organization, whose leaders had to use some anticapitalist slogans to satisfy the more radical sections remaining in its ranks.)

The balance was further altered by the Second World War and the advance of the Soviet tanks to the Elbe. Basking in the glory of the heroic Red Army and of their own role in the Resistance, the Communist parties became the principal forces of the Left in France and Italy. The antifascist alliance between Communists and Socialists did not last long. The cold war made the split sharper than ever. The Communist parties aligned themselves with the Soviet bloc. The Socialists, broadly speaking, chose NATO, the American leadership, and the capitalist world. The name was the same, but social democracy had changed nature. Whether the Socialist parties dropped all links with Marxism, as the German Social Democrats did at Bad Godesberg in 1959, or kept a reference to "the common ownership of the means of production, distribution and exchange," as the British Labour party did, outside electoral rhetoric nobody would have described Helmut Schmidt or Neil Kinnock, for that matter, as a mortal threat to capitalist profits and property.

An argument is sometimes best seen by being pushed to the absurd. When under the influence of the economic crisis, of the Tory offensive, and of the pressure of its own rank and file, the Labour party looked for a time as though it might become much more radical, its most right-wing members breaking away. They joined forces with the Liberal party to form the Alliance, although in negotiating its program, the newcomers revealed themselves as being to the right of their new partners. Indeed, in economic and foreign policy, they seemed curiously close to Margaret Thatcher. And what was the name of this new splinter group? It was called the Social Democratic party (SDP). An organization or a creed, however, should not be judged by its caricature. It is not enough to contribute to the survival of capitalism to be a Social Democrat. As we shall see watching the French Socialists under François Mitterrand, a party does not become Social Democratic through self-proclamation. Taking Britain, Germany, or Scandinavia as an example, the model requires the existence of at least a strong labor movement that is well organized in trade unions and a direct link, preferably organic, between these unions and the political party. (In Britain, for example, the Labour party was set up by the trade unions.) Marxists may accuse Social Democrats of betraying the interests of the workers as a class-for-itself and selling their historical birthright for a mess of porridge; a Social Democratic party to be successful must be seen as linked with the workers and defending their immediate interests.

In this context, the original suggestion of an Italian Communist party well advanced on the social-democratic road sounds more ambiguous than ever. Which social democracy? When Berlinguer outlined his vision of a "historical compromise" between organized labor and advanced capitalism, in 1973 and 1974, the party still seemed comparable with the German Social Democrats of pre-1914 vintage: a mass party, with strong roots in the labor movement and at least a verbal commitment to the radical transformation of society. By now, it looks much closer to the SPD of today—in fact, its chosen partner. Euroleft, a regrouping that extends well beyond the Communist ranks, and no longer Eurocommunism, is the latest European fashion.

The snag with Berlinguer's "historical compromise" is that it missed its appointment with history. By the time it was offered, it was already obsolete. For, at the heart of postwar politics during the period of prosperity, which ensured the success and spread of social democracy, there was an unwritten convention, an implicit compromise. Its basic premise was that the labor movement would not attack the fundamental features of the established social order and the omnipotence of employers at the work place; except after 1968, when it temporarily got out of hand, the movement never really questioned the basic organization of work and the prevailing division of labor. Everything else was negotiable on a national or local scale in a give-and-take, depending on the balance of forces. As part of the bargain, wage and salary earners were given a share of the surplus, which was growing each year as a result of the fast-rising productivity. A striking feature of the period was the expansion of the indirect wage as the state exended its role in the reproduction and qualification of the work force. Family allowances and similar benefits spread in some countries. The educational bill skyrocketed as the second half of the twentieth century tried to do for secondary, and to some extent higher, education what the second half of the nineteenth century had done for primary schools. The feeling of security increased with important improvements in the insurance against old age, ill-health, and unemployment. The working people even gained an impression of security of tenure as some restrictions were put on the employer's right to hire and fire.

Although varying from country to country, the welfare state performed a more important function in Western Europe than it did in the United States as an instrument of economic management, social organization, and political integration. True, in retrospective glow one should not paint in imaginary colors a society gaining mastery over its own fate. Nowhere did the welfare state eliminate fundamental inequalities; its redistributive function was much less important than its propagandists claimed, while the impersonal, distant, and bureaucratic running of its services made it easier to attack the welfare state as an invading Leviathan when the tide turned and, with it, the ideological requirements. But let us not be led astray by the prevailing contrary mood. The postwar changes in living standards, including the extension of welfare, were perceived by the bulk of the population as an improvement. The success of social democracy was not accidental.

Keynesian fine-tuning may have made the capitalist engine run more smoothly, but it had not eliminated its fundamental flaws. By the time Berlinguer was making his offer, the declining rate of profit had changed the financial mood, while the slackening pace of growth and rising unemployment were upsetting the whole equation. Capitalists were clamoring for the carving up of the profitable parts of the property accumulated by the state, for the dismantling of the welfare services in order to cut taxes, and for the elimination of safety nets in order to increase the "flexibility" of labor. Far from proposing new concessions and benefits, the capitalist establishment was trying to win back those it had been induced or forced to grant in the past. Confrontation, not compromise, was now on its agenda.

The new era had not been ushered in unanimously and without opposition. The old *modus vivendi* was too precious to be discarded lightly. The British Establishment, as is often the case, projected openly its inner dissensions. In the late 1970s, through public pronouncements, parliamentary debates, and lengthy editorials, you could piece together the following dialogue: Things cannot go on as they are; the consensus is no longer a paying proposition, and we must change at once the rules of the game—so were arguing, with quotations from Milton Friedman and the monetarist book, the new hard-liners, having picked Thatcher, the "Iron Lady," as their leader. Beware of the consequences; you start by playing at class struggle and find yourself involved in it in earnest with "reasonable" labor leaders who are no longer able to soften the blows—so replied the more patrician, nostalgic upholders of the status quo, dismissed contemptuously by the Thatcherites as the "wets." For national and international reasons, came the reply, we can no longer afford arrangements with even "reasonable" labor leaders. We must prepare for the conflict with the labor movement, break its backbone, and then see whether another compromise is possible or necessary. The choice of Reagan and Thatcher showed who carried the day, while their electoral victories suggested that Social Democratic, or even plain Democratic, solutions no longer appeared adequate to the general public.

It takes two to tango. The European Left suddenly found itself ridiculous, continuing the routine steps on its own. For years it had been so simple. The two main performers were moving in apparent harmony as if nothing of substance stood between them. And now, all at once, the harmony was broken. How can you keep on being inspired by community and compromise when the other side extols the virtue of greed and the survival of the fittest? How can you preach and practice class collaboration when yesterday's partner has become an open enemy? The Left thus discovered that, for a variety of reasons, it was in a very uncomfortable political position.

On the face of it, things should have been otherwise. For the first time since the Second World War, mass unemployment made it possible to present capitalism as a system unable to exploit its own achievements. It can invent computers that are capable of most complex calculations and produce robots that can stand in for people in increasingly sophisticated tasks. These blessings, however, are a curse in disguise, since the higher the productivity, the higher the unemployment. To mount such a counteroffensive, the Left would have had to look beyond the confines of existing society. It needed a radical alternative both realistic and credible. It offered nothing of the sort. Sticking to the system, it was losing the argument because within its frontiers, logic was on the other side: Things obviously could not continue as before.

Driven to the defensive, the Left discovered its shortcomings. The structure, organization, and strategy of the labor unions had not kept pace with the social upheaval and, therefore, with their potential membership. Let us look in succession at the three waves that swelled the ranks of the labor force. Whether one looks at numbers or positions of power, immigrant workers are not found in the trade unions in a proportion corresponding to their presence in the army of labor, and xenophobia or racism is not entirely unconnected with this state of affairs. Not that the unions can be accused of practicing discrimination; but workers, too, are infected by prevailing prejudices, and the unions, like the left-wing parties, have not always dared to attack them head on. Unskilled and semiskilled workers are thus underrepresented in the unions. The Italian exception confirms this rule. In Italy, where the assembly lines in Turin were serviced by migrants from the far south—Calabria and Sicily—the movement of the 1960s and early 1970s was particularly articulate and original. The newly elected factory delegates often spoke with an egalitarian and southern accent. Elsewhere in Western Europe, the unions remained dominated by the traditional "labor aristocracy"—native, skilled, and male.

Feminism has much progress to make in the work place as well as in the home. It is fair to argue that with women often doing two jobs, it is not easy to induce them to take a union task on top of it. It is equally just to point out that labor unions have shown little imagination in persuading women that they defend their real interests and little enthusiasm in offering women important positions at all levels in proportional rather than token capacity. Here, too, prejudices are not totally absent. The underrepresentation of women affects the third wave—the white-collar workers. While teachers and, to a lesser degree, public employees are, on the whole, well unionized in Europe, the same is not true of shop assistants and office workers in private firms. What is at stake here is not only the ratio of union members to the total wage-earning population, which is declining all over. The way in which these often seemingly conflicting interests are being reconciled has a much bigger political significance. Do labor and political leaders just add up various political discontents, promising the unskilled lower differentials and the professionals, the preservation of their privileged position? They may, proceeding like that, occasionally win an election. They have no chance even to begin reshaping society. For this, they need a homogeneous bloc capable of lasting common action. These leaders, therefore, must reconcile only what can be reconciled in a global project, attract through this common vision, and not hesitate to antagonize those whose place is really on the other side of the fence. In the absence of a global strategy, the Western labor movement is, for the time being, unable to embark on this second road.

Finally, there is the debilitating effect of the economic crisis itself. Its classic effect to begin with. Unemployment and stagnating living standards may at some stage bring discontent to the breaking point and lead to rebellion. Their first result, and one that usually lasts quite a time, is to demoralize the workers, discourage the rank and file, and put the movement on the defensive; fear of losing one's job is not an incentive for militancy. On this occasion, the crisis also has a harmful influence on the level of debate within the labor movement. In the 1960s, some union members began to look beyond the traditional problems of wages, to ask questions about the fragmentary nature of their work and about the disciplinary rather than technical reasons for the hierarchical division of labor. In short, they started questioning the very logic of the system. It is this still very tentative debate that was stifled by the new economic climate. When millions of people are seeking jobs, any job, it is allegedly not the moment to ponder the alienating nature of labor. Last but not least, the crisis speeded up trends that were already weakening labor unions: the growing share of women in the work force, the disproportionate expansion of the services, and the drastic reduction of the big battalions of organized labor in heavy engineering and the automobile industry, in the shipyards and the mines.

The weakening of labor was opening up new possibilities for capital, offering opportunities to use legislation in order to deprive unions of established means to organize or picket and their members of rights and guarantees they thought they had acquired forever, such as the minimum wage or the protection of real income through price-indexed wages. It also gave the rulers an occasion to divide—to drive a wedge between natives and foreigners, the skilled and the unskilled, those working in the "protected" state sector and those in the "competitive" private one, the employed and unemployed, and so on.

Considering the advantages accumulated quickly by one side and the handicaps by the other, labor's capacity for resistance is rather surprising. Here again Britain provides an instructive example, with its miners' strike in 1984 and 1985. The Tories picked their target and prepared their plans even before getting into office.[2] Once there, they boosted coal stocks, altered legislation to paralyze strikers' roving pickets, and reorganized the police to have mobile squads at its own command. When the conflict began, the government could rely on the mass media, which was almost unanimously united against the strikers; on the full majesty of the law, thanks to which, say, the funds of the National Union of Miners (NUM) could be seized; on repressive forces of unprecedented size; and on unlimited financial resources. The cost of the stoppage to the nation had to be calculated in billions of pounds, but no price was too high to teach the labor movement as a whole a lesson. Incidentally, those who witnessed such passionate determination on the part of the Tories and still claim that class struggle is a Red invention must be, to put it politely, color blind.

Faced with this armada, the miners had the official token support of only their confederation, the Trades Union Congress (TUC), and of a Labour party leadership busy studying opinion polls to discover what it all meant in electoral terms. That under those circumstances—backed only by their own fighting spirit, the sacrifices made by their families, and the deep and active sympathy of large sections of the working population—the miners could have fought for over a year is almost miraculous. There is no denying the result. The miners were beaten, well beaten. The precedent, however, proved perilous, and the lesson is not exactly what Thatcher and her inspirers wanted to teach. If, almost single-handedly, split from the start, and in an essentially rear-guard action, the miners could have resisted for so long, what would the labor movement do if it rediscovered an offensive strategy and, by the same token, recovered its most precious weapon, a sweeping solidarity that no leadership could afford to ignore. The trial of strength in Europe is not over yet. But let us not anticipate.

3

Seeds of Defeat in Victory

There are general laws and the peculiar, specific cases in which these tendencies materialize. Nineteenth-century France was treated as the best political laboratory because the political conflicts of that era were pushed there to their logical paroxysm. France in the second half of the twentieth century is equally instructive for the understanding of our age.

At mid-century, the parliamentary Fourth Republic—with its governmental merry-go-round, its ministerial musical chairs, its Communist party capturing one quarter of the country's vote and thus revealing the strength of public discontent—was clearly not in a position to fulfill at the same time its two imperative tasks: to modernize the economy in order to allow it to face European competition within the projected Common Market (with which, on its own, France seemed to be able to cope) and to extricate itself from direct involvement in a colonial empire. The war in Algeria offered the possibility of changing the institutions, together with the number of the Republic. As Cincinnatus at Colombey-les-deux-Eglises, General de Gaulle was getting restless with his plow. The military coup of May 1958 gave him the eagerly awaited occasion to return to power and to do so on his own terms.

One of his first moves was to impose a new constitution. This charter, which, with a few amendments, is still valid, is a hybrid document, neither purely parliamentary nor fully presidential, although heavily tilted in the latter direction. If the Fourth Republic was seen as the reign of the parliamentary assembly and of political parties, with the president a mere figurehead, the Fifth introduced the rule of the executive, with the parliamentary assembly reduced to the rank of a rubber stamp: The president can, if he deems that the institutions of the Republic are threatened, take all powers provisionally into his hands.[1] He can seek the verdict of the people at any time either through a referendum or through an election, resigning or dissolving the National Assembly at the moment of his choice. The executive has the initiative in introducing legislation. And we could continue the list of biases.

All this was not quite enough. The French ruling class had resigned itself to de Gaulle when it had to. Once he had done his Algerian job, it might have been tempted to get rid of the awkward defender of its long-term interests. The general, a master tactician, was ready with a parade. In 1962, the Algerian war finally over—it had lasted for four years under the new regime and eight years altogether—he asked the French people to approve in a referendum his proposal that the president no longer be chosen indirectly by an electoral college of 80,000 notables (senators, deputies, although essentially local councilors) but directly by universal suffrage. Despite the opposition of most politicians, the French electorate, as might have been expected, endorsed his demand, thus shifting considerably the legal balance of power.

The president, elected for seven years, would soon be able to tell the deputies, elected for five years, that individually they were just a fraction of the popular will, whereas he on his own represented the nation as a whole.

The first presidential election by universal suffrage was staged in 1965.[2] There was no great rush of candidates on the Left to measure up to the legendary hero. This was the great, historical opportunity for the man who will figure prominently throughout this book. François Mitterrand, politically a loner, jumped on the stage and imposed himself as *the* candidate of a united Left. He succeeded beyond expectations. General de Gaulle was not elected on the first ballot, which requires an absolute majority, and in the second, in which only the two top candidates are left to fight it out, he outpointed Mitterrand by 54.8 percent to 45.2 percent, a splendid score by normal presidential standards, although mediocre for a man who had captured 80 percent of the vote on his return to power. The elected monarch no longer looked of divine right.

Concentrated power is a source of strength and of fragility. The ruler, not bothered by intermediaries, has few warning signals, and each major conflict tends to become a national drama. In the late 1960s, student rebellions spread throughout the world from Berkeley to Tokyo. Only in France and, later, in Italy did the student rebellion get a response, however complex and ambiguous, from the workers. Only in Paris, after the whole country had been paralyzed for weeks by the biggest general strike in the nation's history, was the question of power at the top put on the political agenda. Admittedly, once he had grasped that the Communists, far from guiding the upheaval, had their eyes fixed on the electoral horizon and were frightened by a movement they did not control, de Gaulle recovered his poise and, after a period of interregnum, his position. Since nothing fails like failure, the Left paid a price for "sins" that it had not committed; the Communists in particular, champions of order during the events, were unfairly depicted afterward as "car burners," supreme crime in Western society. And since nothing succeeds like success, in a climate where fear had replaced hope, the Gaullists were returned to office in a snap election with a bigger majority than in 1967—"a majority of funk," to borrow de Gaulle's own contemptuous expression.

On the face of it, everything was back to normal. In fact, nothing was going to be the same as before. Yesterday's savior had become for the Establishment a potential peril. Ostensibly, de Gaulle tried to restore his magic, asking the French people to approve a project—on decentralization—by a referendum. Actually, he was consciously committing political suicide. Defeated, in April 1969, he went back to Colombey. His place was taken by an ungrateful lieutenant. The ruling class no longer needed a Napoleonic figure parading on the world stage and challenging the United States. It wanted a Louis Philippe, a bourgeois king and protector of the propertied classes, and found one in Georges Pompidou, a former chairman of the Rothschild bank and of not-too-distant peasant stock. The transition involved no real risks because, with the Left divided and splintered, the result was a foregone conclusion: Pompidou was duly elected president.

The real problem, however, was how to get this extraordinary interlude forgotten, transfigured, or even exploited. During the weeks when the country was at a standstill, the students were not the only ones to ponder their function in society. In the idle factories and offices, millions of men and women asked questions about the military discipline of work, the hierarchical order of society, and, ultimately, the meaning of their lives. It was vital to put a stop to such nonsense and to convince those who caught a glimpse of another life that it was no more than a mirage. There was here a clear division of labor. The Right was to urge people to make money, while it was ensuring law and order. Pompidou presided over the last fling of rapid expansion as though he had a foreboding that the time was running out for the system as well as for him. It was up to the Left, naturally enough, to channel revolutionary currents into parliamentary waters.

The Communist party was particularly shaken. It had emerged from the crisis with its revolutionary uniform in tatters. It is irrelevant whether the situation had or had not been potentially revolutionary. In a country paralyzed by the biggest strike ever, the party, whose raison d'être was to push the movement as far as it would go, had done everything in its power to keep it in check, and this maneuver had been obvious. On the morrow, it was important for the party leaders to dismiss all those to the left of the Communist party as dangerous anarchists and to convince their own supporters that if the revolutionary dream was utopian, quite a radical transformation might be achieved by parliamentary means. This posture can be understood only in the context of the party's evolution.

The first decade after the Second World War was one of spectacular rise for the French Communist party. Helped by its record in the Resistance, by the militancy of its members, and by its control of the biggest labor union, the General Confederation of Labour (CGT), the Communist party became the principal force of the French Left. Its ascendancy among workers and intellectuals was greatly assisted by its contrast with the Socialist party, or French Section of the Workers International (SFIO), which under the leadership of Guy Mollet had opted for open collaboration with the

Right and the conduct of a colonial war. The year 1956 may be taken as a turning point in the fortunes of the Communist party. It was not only the year of Nikita Khrushchev's speech and of the collapse of Stalin's statue. The invasion of Hungary suddenly showed that the Red Army could turn its guns against the workers. In France, it was the second year of the Algerian conflict, and the Communist party, although not in office, voted the special powers allowing the Mollet government to carry on the war. This and, more generally, the Communist refusal to back the Algerian insurrection antagonized a whole generation of young militants.[3] Two years later came the first big electoral setback. De Gaulle's victory cost the Communist party dearly. The general appealed to an important fringe that had voted for the party as a protest against the "rotten system" and was attracted by his "national grandeur." The Communist vote dropped in 1958 from 25.3 percent to 18.9 percent, and, although it subsequently picked up, it never quite recovered. To put it roughly, until then it had hovered around 25 percent of the French electorate; thereafter, around 20 percent.

However serious, the drop was not tragic; it had its compensations for the leadership. It justified the stifling of any serious debate after the shock of the revelations about Stalin: During an emergency, you cannot afford any relaxation of internal discipline! It also allowed the Communists to get out of their ghetto. Their prestige was partly based on the fact that they were not mixed up in any shady deals. They were outsiders, the untouchables of parliamentary politics whose votes counted negatively, to bring down a government, not positively, to vote its investiture. All this was possible because under the Fourth Republic, elections were held under a system of proportional representation, each party getting a number of deputies in a region proportionate to the number of votes it obtained there; deals were struck afterward in parliament. The Fifth Republic changed it all by reverting to a system of majority voting in small, single-member constituencies, quite close to the British and American model, except that there are two ballots in France. In the first, which requires that a candidate win an absolute majority to get elected, the voters can pick and choose, expressing their preferences. In the second, in which a simple majority is enough, they tend to eliminate. The electoral mechanism has a polarizing effect, and, traditionally, one candidate of the Right and one of the Left face each other in the second round in most constituencies. The same thing happened in the early 1960s. As soon as the Gaullists seemed to be uniting the Right, electoral logic proved stronger than rabid anti-Communism. By 1962, Mollet's Socialists were making electoral deals with the Communists. It was to preserve this unity that three years later, the Communist party accepted Mitterrand as its presidential candidate. In a way, it paid off. In 1967, the newly created Federation of the Democratic and Socialist Left (FGDS), headed by Mitterrand but numerically dominated by the Socialists, made an electoral pact with the Communists on the national scale, and together they almost won that year's parliamentary poll. They opened the new year by signing a common platform. With the goal in sight and their eyes fixed on the electoral horizon, no wonder they missed what was happening below and were taken completely aback by the upheaval. Their plans shattered, they had to pick up the dispersed pieces and start all over again.

Was not the Communist party then in a position to take advantage of the new situation? With a generous provision of ifs, the rewriting of history is an easy exercise. After all, if the French Communist party had been what it was supposed to be, the May events would have taken a radically different course. All that can be safely said is that afterward, French Communists missed an opportunity to gain on the French Left the dominant position that their Italian comrades occupy in their own country. With the Socialists in an even worse plight than themselves, the possibility for the Communists was there. To seize it, they should have, like the Italians, broken their links with the Soviet Union and destroyed their Stalinist image more rapidly in order to appear as the only genuine left-wing alternative. Taking half-measures half-heartedly, the party was unable to win a hegemonic position. At the same time, to preserve its parliamentary strategy, it required a partner, and far from squeezing the Socialists, it was, ironically in retrospect, ready to help them survive.

The non-Communist Left at the time needed all the help it could get. Its influence was at its nadir.[4] It was torn apart and bickering. Mitterrand had been pushed aside as the man whose strategy and conduct had been responsible for this predicament. But he was no man to lie low for long. By mid-1971, at the Epinay congress, designed to turn Mollet's old SFIO into a new, broader Socialist party, he managed to combine his debut in the Socialist arena with a successful takeover bid. His faithful friends from the Convention of Republican Institutions (CIR)[5] being too few—97 out of 957 delegates—to ensure his ascendancy, he made a deal with the Center of Socialist Study, Research, and Education (CERES) of Jean-Pierre Chevènement, on the left of the party, as well as with Gaston Defferre and Pierre Mauroy, reputed to be on its right. Their coalition won the day by the narrowest of margins: 43,926 mandates to 41,750 mandates, with 3,925 abstentions. It was enough for Mitterrand to become the party's first secretary.

For the first time, the solitary politician had an instrument at his disposal. He was to use it "to redress the balance of the French Left," that is to say, to strengthen the relative weight of his new party. Yet to do so, he had to show that it was really a party of the Left, that the days of alleged nonalignment of the so-called third force and, in practice, of collaboration with the Right were over. Hence, the first task was to reach an agreement with the Communist party. Within a year, in June 1972, the two parties signed their famous Common Program.

In one sense, that program was the respectful Left's answer to the May movement. Whereas the May movement, in a rather utopian way, was dreaming of a different society, the Common Program was suggesting how to improve the existing one by extending the public sector and by granting more power to the labor unions in the factory, more money and better services to the working people, and so on. According to its authors, it did not propose to build a socialist society at once, merely to prepare the way for one. Actually, it was a contradictory document: too timid to lay the foundation of another system, yet too ambitious to fit into the exising framework and to be accepted by the powers that be. It did not matter very much. This blueprint, based on a very fast rate of growth, was soon to be overtaken by the economic crisis. Its primary purpose was to reunite the French Left and to relaunch it on the way to electoral victory.

The first test, the parliamentary poll of 1973, coming up too soon for success, the Left was at once given another chance. The death of Pompidou in 1974 precipitated an unexpected presidential election, which probably marked the climax of the united Left. Mitterrand was now its natural candidate, and he carried the campaign with enthusiasm. The Communists went out of their way to facilitate his performance. Because of their past, they still frightened waverers, so they proclaimed on their own that in case of victory, they would not ask for the ministries of defense, interior, and foreign affairs, the very portfolios that de Gaulle had argued should not be entrusted to Communists. It was all to no avail. Mitterrand missed victory by a cat's whiskers. Although barely 1 percent separated the two candidates, Valéry Giscard d'Estaing became France's new president.

It was both a victory for the Right and a funeral for Gaullism. Pompidou had acted as a transition. Giscard, although he had served under de Gaulle for years, belonged to the other, less authoritarian tradition of French conservatism. Distinguished, technocratic, and a pure product of the Establishment, the newcomer began by pretending that France wanted to be governed "in the middle." Lowering the voting age to eighteen and allowing abortion to be legalized, he himself first donned the mantle of a reformer. His misfortune was that his reign coincided with the spread of the economic crisis. When it came to serious business, his own and his electorate's wallet was well on the Right. The reformist or "centrist" illusions did not last.

The economic crisis, with "stagflation" and rising unemployment as its companions, was affecting the political as well as the ideological equation, and it was no longer safe to leave the policing to the Left alone. The hopes born and the discontent revealed in 1968 had been successfully diverted, not eliminated altogether. The campuses, the factories, the offices were still potentially explosive. New forms of dissent had since made their entry on the French social scene—ecology rather timidly, feminism with greater vigor. However, these protest movements did not really threaten the political edifice because they were isolated, sporadic, and separate. The danger was that economic discontent might link them together and provide the lead for a frontal assault. With growth, one of its pillars—the other being pluralism—removed, the edifice was really threatened. It was urgent to convince the potential attackers that while to rebel might be just, to prepare a general strategy, a global project, was politically absurd and morally suicidal; it could only lead to the "concentrationary" universe, to the gulag. And the efficiency with which the social need, if one may say so, required for the defense of the existing order was fulfilled is staggering.

Let there be no misunderstanding. I am not alleging that in the advanced capitalist countries, there is a secret mastermind manipulating propaganda at all levels, that scribblers who do not obey are sent to Siberia, while those who do live with a golden collar round their necks. One is tempted to say that there is almost no need to bribe or twist, although considering some of the salaries, the word *almost* is not superfluous. More seriously, the point is that to censor, to bribe, or to bully is not a sign of strength. The real art is, without the ostensible tools of coercion, to produce the right message at just the right time—for example, to make the French "discover" Soviet concentration camps in 1975 and to turn this belated discovery into a major ideological and political event, announced by the "intellectual" magazines, and echoed by the press, radio, and television, drumming into the ears of the young the gloomy slogan that all revolutions are condemned to tyranny and there is just no alternative. Madison Avenue experts must have been green with envy watching how the publication of Aleksandr Solzhenitsyn's three-volume *Gulag Archipelago* was built in Paris into a successful operation whose code name was *nouveaux philosophes*.

It was still too early at that time to let learned professors like Raymond Aron—say, a superior Daniel Bell—lecture young people about the evils of Communism or the inevitability of tyrannical endings to Promethean efforts. They would have been yawned off as biased fuddy-duddies. Required for the role were members of the younger generation, preferably with a real or an imaginary connection with the May movement, possibly ex-Maoists ready to preach their new creed with the same passion they had shown when tossing their little red book a few years earlier. They had hoped, believed, and practiced, the argument ran, and now they are back from their journey with the bitter truth. Recantation as a new form of revelation. And if you do not trust them, by their side is Solzhenitsyn, a man who has personally paid the price and who now bears witness for your enlightenment.

Another misunderstanding must be avoided in this connection. Solzhenitsyn's reactionary prophetic pronouncements in exile do not destroy the value of his testimony. The silence of large sections of the Western Left at the height of Stalin's purges was shameful, and their reconsideration is better late than never. A debate without any taboos on the roots, extent, and unfinished consequences of this tragedy could only do the Left some good, provided that it were carried out rationally and within a historical context. The new philosophers simply revived the old incantation in reverse. Marx was now the begetter of barbed wire, as he had once been the holy father of deified Stalin or Mao. The argument was equally primitive, equally Stalinist, only with the target reversed. Yesterday, the Soviet Union (or China) could do no wrong and Western imperialism, no right. Today, Russia was the "empire of evil," and anything Western was right by comparison.[6]

The name is clearly a misnomer. There is no philosophy in this very old show. Young Frenchmen on the make tried to repeat, on an infinitely lower level, the ceremony of revulsion performed in the English-speaking countries shortly after the Second World War and illustrated by a book that had its moment of fame, eloquently titled *The God that Failed*.[7] Those who held even this original act of collective recantation as being half-earnest and half-ridiculous should know that the repeat was the parody of a farce: a sprinkling of ideas pinched from Karl Popper, a dash of quotations from Friedrich von Hayek, a few statistics from recent books about Soviet repression published in English, mixed with a good portion of Solzhenitsyn. The French contribution to this mixture was empty rhetoric. As one critic put it, "a kamasutra of recantation in 100 postures."[8]

What else was particularly French about this whole exercise? The ease and arrogance of intellectuals jumping from bandwagon to bandwagon. They had no moment to spare for contrition, regret, or reflection. They were literally claiming laurels for their past blunders. Their former stupidity was a guarantee of their present wisdom. They had to be treated as *voyants,* or "seers," for the very reason that they had been blind in the past. The intellectual contribution of new philosophy was not overwhelming, and the names of its practitioners may not be vital for posterity. But the political success of the operation went beyond the hopes of its sponsors. Thanks to a repetitive campaign of the media, the operation prepared the terrain for a complete revaluation of values. Not in Germany, Britain, or the United States, but in France, where the domination of the ruling ideology until then had not been complete, and in Paris, where a few years earlier there had been a question of imagination seizing power. In the same capital, it soon would be possible to argue, pretending to belong to the Left, say, that there is no reason to feel guilty about colonialism, considering what happened in the former colonies after they gained their independence. Later still, a new refrain would repeat that Marxism is the source of all evil and the market, the fountain of freedom. The new philosophers really did prepare the ground for Reagan and Rambo, for a bout of McCarthyism without the witch hunts, for a return to the Dark Ages in the *cité lumière,* and this fundamental change in the ideological climate would inevitably affect the Left once it reached office.

This was still to come. In the meantime, the Socialists were not entirely displeased with a trend that was weakening their partners and, nevertheless, rivals. The Communists, victims of this campaign because of their contradictory antics, unwittingly contributed to its success. The honeymoon between the two parties of the Left did not last long after the 1974 election, nor is this surprising. Each had entered the match determined to turn it to its own advantage. Mitterrand told the Socialist International in Vienna almost at once that his objective was to win about 3 million Communist votes. Georges Marchais told a secret session of the Central Committee that the Socialists were inveterate Social Democrats and not to be trusted.[9] The Communists used to brag in the past about "plucking the Socialist chicken" (*plumer la volaille socialiste*). They now found themselves in the posture of the skinned rather than the skinner.

In order to restore their position and capture a portion of the vote of the so-called new middle classes, the Communists decided to brush up their image. They did it much too late and advanced so clumsily as to lose on both the swings and the roundabouts. Having condemned the Soviet invasion of Czechoslovakia, they then resigned themselves quite comfortably to the "normalization" in that country. They now embarked on an important revision of their theoretical and practical arsenal—dropping, as we saw, the concept of the dictatorship of the proletariat and switching from the rejection to the approval of the French nuclear deterrent—but these reappraisals were neither preceded by a genuine debate nor accompanied by serious explanations. Marchais having discovered a passion and a certain clownish talent for the little screen, the party members were more likely to learn the latest twist in the party line by switching on their television sets than by attending meetings in their cells. Rumor had it that if the party were suddenly to abandon "democratic centralism" and hand all the power back to the rank and file, this too would be announced in a ukase handed down from above and read by Marchais during a talk show.

It was not just a problem of method and manners. The Communists were also painfully learning a rule: Within a coalition linked by a moderate program, the party with the more moderate reputation is likely to gain more. The Communists were much stronger than their Socialist partners on the shop floor and in the streets, and they had shown it, if only negatively, in 1968. They were now reduced to an electoral confrontation in which they were the losers. The Socialists, using the language of May 1968 and often turning the Communists on their Left, had in the meantime won the battle for the teachers and technicians, for the "new strata" and professional intelligentsia. Logic was driving the Communist party into the position prescribed by its ally, that of junior partner, and since it rejected that role, the two parties were on a collision course.

An up-dating of the Common Program served as the pretext for the breach. With the Left poised for electoral victory, a thorough reassessment of the program drawn on the assumption of a fast rate of growth was long overdue. What the Left offered its public in 1977 was the semblance of a debate. The only leader to treat the new environment in earnest was the clever, if versatile, Michel Rocard. Small, nervous, often a convincing speaker with a machine-gun delivery, Rocard had been only a few years earlier the leader of a smaller and much more radical socialist group, the Unified Socialist party (PSU). On joining the bigger and more moderate Socialist party, he shifted so fast as to find himself on its right, or social-democratic, wing. Consistent with his new vision, he argued that the Left had to revise its objectives downward, since it no longer had the means to carry out its original goals. (It did not cross his mind to maintain, as the former Rocard would have done, that the Left should radicalize its means to achieve the original ends.) At the other extreme, the Communists were projecting a Japanese rate of growth, as though nothing had happened and the outside world did not really exist. In the middle stood Mitterrand, who, like all wise politicians, knew that electoral battles are not won by withdrawing pledges on the eve of the poll: The less said about the subject, the better; and the matter itself can be solved after victory.

The Communists' case against the Socialists was not only weakened by the surprisingly belated discovery that their partners were reformers and not "revolutionaries," but also spoiled by the absence of coherent proposals to adapt the Common Program to fit entirely new circumstances, of suggestions about how a Japanese rate of expansion could be reconciled with balanced foreign payments, and of measures designed to alter institutions and to mobilize people in order to be able to resist the inevitable attacks of the "gnomes of Zurich" and their Parisian banking brethren. All was to be solved by a magic number of nationalizations. "Fifteen nationalizations represent class collaboration, while sixteen make a revolution," Mitterrand could note caustically in his diary on an earlier occasion.[10] No wonder there was a sneaking suspicion in the left-wing electorate that the Communists would rather lose than play second fiddle in a Socialist government during a period of crisis. This impression was confirmed by the beaming, almost hilarious face of Marchais on television, belying his sorrowful platitudes, on March 19, 1978, in the hour of defeat. For months, the results of local elections—cantonal and then municipal—opinion polls, and all other indicators had been pointing to an inevitable defeat of the Right. Then came the split. The divided Left did not look credible or deserving to win. It ultimately succeeded in snatching defeat from the jaws of parliamentary victory.

It was not a defeat like all the previous ones. This time, with the Communists refusing to accept their appointed place, the whole strategy of left-wing unity was in ruins. In the Socialist party, in non-Communist unions, and in friendly publications, the question of Mitterrand's future was raised openly. The so-called Second Left, which we shall meet again in this story but which can be summed up here, with hindsight, as using the language of the New Left better to sell a social-democratic line, was praising Mitterrand for his past achievements and urging him to step down so that its hero, Rocard, could lead the Socialists to more successful battles without the Communists as allies.[11] The *deuxième gauche* did not talk of moving to the center, since it was still too early to revive the idea of a third force; yet in a country where the non-Communist Left was capturing 25 percent of the total vote, this clearly was the arithmetic logic.

Once he recovered from the shock of shattered hopes, Mitterrand had no difficulty in winning the battle within his party. With the help of the old guard from the CIR and the younger leaders known as sabras who had rallied around him since the Epinay congress—including Lionel Jospin, Laurent Fabius, Jacques Attali, and Paul Quilès—he reduced Rocard to a marginal position in the Socialist party and forced him to make his potential presidential candidacy dependent on Mitterrand's own decision. Significantly, he won this trial of strength by sticking to his line. The Left's only chance of victory, he maintained, was with Communist support. If the Communists give up unity, the Socialists must appear as "unifiers for two." This, incidentally, is also the only way to win Communist votes. If it has not happened yet, it soon will.

Such a line had implications for the campaign. Mitterrand, whose credibility, almost nil after the defeat, rose gradually as the 1981 presidential poll drew near, had to appear as being true to himself. The Socialist program, the candidate's own 110 proposals—including nationalizations, a big raise for low wages, and social benefits—was a direct heir to the Common Program signed with the Communists. At the same time, in his own campaign, Mitterrand sounded moderate, like the calm strength, the "tranquil force" of the posters showing him against the bucolic background of a French village with a church steeple and all. Those who claim that the Socialists promised the French a "break with capitalism" are right, since *la rupture* was a fashionable word at the time in the Socialist vocabulary. Those who claim that Mitterrand promised them peace, quiet, and continuity are also right. (And if, as in the story of the refereeing rabbi who successively announced that both sides in the conflict were right, you claim that they could not both be right at the same time, you are right, too.)

Whether radicalism or moderation carried more weight is really irrelevant. The Left did not win the 1981 election. The Right lost it because it was divided, the neo-Gaullist supporters of Jacques Chirac refusing to be squeezed by the more classical conservatives standing behind Giscard, the outgoing president. It lost because it was clearly the twilight of a reign, scandals like the one over diamonds offered to Giscard by Jean-Bedel Bokassa, the cruel "emperor" of Central Africa, contributing simply to the feeling of decay. It lost, above all, because it revealed itself to be unable to cope with the economic crisis, as was shown by the rise of unemployment from 2.8 percent of the labor force in 1974 to 7.4 percent in 1981. It seemed high time to give the other side a chance. To top it all, the Right waged a campaign of another age, equating the possible victory of the Left with a major national disaster. If it did not quite swear that both private houses and wives would be collectivized, it was only just, just. It described the Socialist party as poor, naïve Little Red Riding Hood bound to be swallowed by the big, bad Communist wolf. The favorite weapon turned out to be a boomerang.

The first ballot confirmed Mitterrand's tactical assumptions. Marchais, whose active campaign across the country quite often gave the impression of being directed mainly against the Socialists, obtained only 15.5 percent of the total vote, by far the lowest percentage scored by the Communist party since the Second World War. To make matters worse, the gap now widened, Mitterrand getting 26 percent. It was a case of Little Red Riding Hood swallowing the wolf. The Right, which had focused its campaign on this issue of Communist strength and Socialist weakness, had no leg to stand on. It felt defeated. Giscard was left to fight it out with the Socialist. In the by-now traditional television duel between the two men, although the smooth, efficient, technocratic machine was still ticking, there was no spirit behind it. Giscard looked like somebody who knows that he is beaten.

And we thus come back to the second ballot, the final suspense, and the dancing at the Bastille.

This short refresher course was not meant to prepare the reader for an examination in recent French history. It was designed and selected to stress three major handicaps of the Left, which were not fully perceived in the hour of glory and the excitement of success, but which must be understood if the subsequent story is to make sense.

The French Left climbed into office without any prior public discussion of the new economic predicament and its political consequences, without any serious and open debate about the respective advantages and disadvantages of protectionism and unrestricted international division of labor, without consulting its supporters about what was to be done when the policy of a Socialist government met the inevitable hostility of its capitalist environment. It was not just a question of clever electioneering, avoiding alarms, and the less said about it the better. Admittedly, it was a piece of rhetoric when, at a party congress in Metz in 1979, Laurent Fabius, the future prime minister, proclaimed peremptorily: "But between the plan and the market, between you and us, Michel Rocard, there is socialism." Yet guided by reflexes conditioned during years of regular growth, the Socialist leaders seemed honestly to believe that a better command of economic levers, coupled with a dose of Keynesian pumping, would relaunch the stalling engine and allow a journey without major obstacles. It apparently did not cross their minds that a Socialist government must govern differently from a capitalist one, must mobilize, because it has no chance of success unless it is carried by a wave of popular support. Faced with the choice that no left-wing government can afford to avoid for long, between radicalization or surrender, the Socialist leaders were totally disarmed.

They were the more disarmed because the ideological climate had altered and was still changing fast. Equality was being turned from virtue into vice. An abstract, undefined state was becoming the only monster to be hated, and freedom was switching sides as it was increasingly identified with buying and selling. We saw the beginning of this operation with the *nouveaux philosophes*. Symbolically, its first climax was reached in June 1980, when thousands of battle-scarred veterans mixed with youngsters to escort Jean-Paul Sartre to the Paris cemetery of Montparnasse. It was the close of a period, the (provisional?) end of the era of commitment. Another chapter in the long saga of intellectual betrayal, of *trahison des clercs,* was being resumed. The seizure of power, emphasized the Italian Marxist Antonio Gramsci, is not only a problem of political takeover. To be real, it must involve the gaining of cultural as well as social and economic hegemony. In France, it was a case of electoral victory combined with ideological defeat.

Last but not least, it was an electoral victory neither preceded nor accompanied by a vast social movement. Although it took them time and effort, the Socialists and Communists did finally convince their supporters after 1968 that if there were no revolutionary shortcuts, life could be altered gradually through the ballot box. The workers were taught not to fight too much, not to strike too often in order to allow their representatives to win seats and obtain in parliament the same results that they would have obtained through their battle. Then in 1978, the parliamentary road, in turn, was shown to be blocked. All hopes were dashed. The miracle at the polls three years later came out of the blue, like pennies from heaven or a lucky draw at the lottery. The French Left did not conquer power; it had electoral victory thrust on it.

How can you talk of passivity, of gifts from above, the reader may object, when you are watching an enthusiastic crowd some 200,000 strong dancing around the Bastille within hours of a successful poll? Granted that history is more complex than any formula. The coming to an end of twenty-three years of uninterrupted conservative rule did arouse passion, and the enthusiasm did open up political possibilities. Yet listen to the revelers more carefully, and draw a comparison with the past.

In June 1936, the only previous occasion when a popular-front coalition of Radicals, Socialists, and Communists had won a general election, French workers were occupying their factories. The frightened property owners, or rather their representatives, had to beg Léon Blum, the Socialist prime minister, to save their skins, and the social order, by granting concessions; and a forty-hour work week or two weeks of vacation with pay for everybody was not negligible fifty years ago. In 1981, the men and women at the Bastille were echoing the slogan from sports stadiums, "On a gagné"—translated as simply "We have won"—and chanting "Mougeotte aux chiottes" and "Elkabbach à la météo"—which are best rendered as "Dan Rather to the shithouse" and "Barbara Walters should do the weather report." Even allowing for the greater importance of the mass media in our time, the difference between the two movements is eloquent. The thirteen years since 1968 had taken their toll. The gist of what the nice people at the Bastille were actually saying was "We have won, they will govern, and François protect us. . . ."

4

One Character in Search of a Role

By the time, well past midnight, when the revelers, drenched but happy, were drifting home, a car stopped in front of number 22 in a small, narrow street of the Latin Quarter, quite close to Notre Dame. With its seventeenth- and eighteenth-century houses restored, the rue de Bièvre, formerly a rag merchants' street and then a shelter for poor North Africans, was an extreme case of gentrification. A couple of Labrador retrievers had to be dragged out of the car, but first, applauded by a few local sympathizers, emerged the hero of the day. The prospective president had had time to ponder his fate during the drive of 150 miles or so from Château-Chinon, his constituency in the Nièvre in central France, to his Parisian home. He—the eternal loser; the horizontal champion; the Poulidor of French politics, from the name of a very popular cyclist eternally second in every race—had finally climbed to the very top. After almost forty years of political activity, after twenty-three full years of strenuous, often seemingly Sisyphean efforts harnessed to one always elusive goal, he had finally grasped it when it looked permanently out of reach. In this, his finest hour, he may well have been looking back at the vagaries of his strangely checkered career.

François Mitterrand was not predestined to be a Socialist

president. As he himself put it, he "was not born on the Left, even less a Socialist," and he "was never a producer of surplus for the benefit of another." Actually, he was born on October 26, 1916, in Jarnac, in the Cognac country, into a family of provincial notables. His father, who had worked as a station master and then in insurance, took over the vinegar-manufacturing business that belonged to his wife's family. The only possible link with the Left in Mitterrand's environment was a distaste for pushing, conquering capital: Mammon was the enemy of God. Yet, to quote Mitterrand again, it was enough to hear how contemptuously "the word 'materialism' was pronounced by these honest people to understand the distance separating them from an intellectual adhesion to socialist ideas." This, if anything, is an understatement. Mitterrand came from a provincial background that was conservative and deeply nationalistic.[1] Nor did the youthful Mitterrand rebel against his environment, either as a schoolboy getting a good classical education at the college of Diocesan Fathers in neighboring Angoulême or later as a student of law and political science in Paris. He did not get involved in the passionate battles that split the Latin Quarter, like Europe itself, in the 1930s. Much later, talking of Franco's Spain, he was to comment regretfully that at the time of the civil war he had been "looking from far, from too far" at this conflict.[2]

One feature in the young man that pointed to the future, beyond the undoubted ambition, was his refusal to yield, his determination never to say die. Like one of those Russian dolls, the *vanka stanka,* which, no matter how many times you pull down will always pop up, Mitterrand was going to move through his political life with an uncanny capacity for recovery, and this resilience was clearly there from the start. Called up for national service in September 1938, he rose to the rank of sergeant in the colonial infantry. Wounded and taken prisoner by the Germans in June 1940, he tried to run away twice and failed. He tried a third time, in December 1941, and was successful. The escape led him logically, although not at once, into the Resistance. Tales spread occasionally by his opponents about Mitterrand's Pétainist past are malicious gossip. He did work for a time in Vichy, the capital of collaboration, in a body dealing with prisoners, but he actually kept the job as a cover on orders from the underground. His Resistance record is unimpeachable. Otherwise, he could not have become, in August 1944, Secretary General for Prisoners, a sort of minister, in the temporary administration set up by Charles de Gaulle on French soil before he brought his provisional government from Algiers.

The underground period led to at least two meetings that were to mark his life. One, through her sister, was with Danielle Gouze, eight years his junior, who was to become his wife. Daughter of a schoolteacher turned headmaster, who was kicked out of his job by the Vichy regime because, among other things, he had refused to draw up a list of Jewish pupils and teachers in his establishment, she really was "born on the Left." Her background was progressive where his was conservative; lay, probably anticlerical, where his was religious.[3] To suggest, as Mitterrand later did in an interview, that "she always finds me much too moderate in my political life" may or may not have been a figure of speech. It is more than likely that, whether on Franco's Spain or on Nicaragua later on, she takes by instinct positions that Mitterrand can reach only through reasoning. Paradoxically, her family represented, in an idealized form, the very image of his future electorate.

The second meeting was a brief encounter with General

de Gaulle, in Algiers in December 1943. Because of some conflicts between two underground networks dealing with prisoners—one Morland's, to give Mitterrand his name from the Resistance; the other, although less representative, having been recommended to the general and headed by his own nephew—this first tête-à-tête was very far from successful. Naturally, the young Resister did not, as a result, lose his respect for the then undisputed leader. There was, however, in his case no sympathy in the etymological sense of the word, no deep devotion, which was to mean so much drama for other non-Communist Resisters when they were soon summoned to choose between the general and the Republic. Mitterrand, although no blind admirer of the Fourth Republic—he actually voted against its constitution—had no such qualms of conscience when separating from de Gaulle, and this attitude also had a bearing on his future.

Thus politics became almost at once the field for his ambition. Dark, handsome, attractive to women, and far from indifferent to their charm, the young man finding his way in postwar Paris did not at once practice law, for which he was qualified, but dabbled in journalism for a time.[4] He was actually gifted for both. He has always been an effective and eloquent, if sometimes overelaborate, orator. He has also been an elegant, highly professional, and brilliant, if occasionally long-winded, writer. Altogether, he has always been better in his use of the spoken or written word as a rapier in political and personal combat than as an instrument of moral incantation, which quite often sounds less than sincere; he is a better polemicist than preacher.

Reluctant to play second fiddle in a vast orchestra, he joined one of those center-left groups that are in France compared to a radish: red on the outside, white on the inside, and always on the side where the bread is buttered.[5]

The one he chose, the Democratic and Socialist Union of the Resistance (UDSR), was so small that at times to form a parliamentary group, it required as allies a number of deputies from black Africa. (One of them, the future president of the Ivory Coast, Félix Houphouët-Boigny, quipped that it was the only place where blacks could watch whites devouring one another.) The UDSR, however, was not too small to have two wings: a Right, headed by René Pleven, the man in charge at the time of the fall of Dien Bien Phu, and a Left, of which Mitterrand rapidly became the chief spokesman.

Center-left would be more accurate, judging by the radicalism of Mitterrand's policies in office. As minister of overseas territories in the Pleven government (July 1950 to February 1951), his boldest idea was to move toward a federation under French auspices. He thought that "colonial society could be transformed by nonviolent means." Experience taught him that "it was in itself violence" and that to remove that violence, one had to get rid of colonial society. "Having understood it, I took some time to admit it."[6] As minister of the interior in the government of Pierre Mendès-France (June 1954 to February 1955), he had a key position at the outbreak of the Algerian insurrection, at dawn on November 1, 1954, which he greeted in a parliamentary speech with the famous refrain "Algeria is France." Granted that many politicians at the time joined in the same chorus. Mitterrand, however, was also minister of state in charge of justice, or Keeper of the Seal, throughout the reign of Guy Mollet (February 1956 to May 1957) and as such was at least co-responsible not only for the Suez expedition, but also for General Jacques Massu's bloody pacification during the notorious Battle of Algiers. So much for the legend of Mitterrand the leftist almost from the start.

Not radical enough for some, he was still too much of a reformer for others. The colonial lobby accused him of "handing black Africa over to the Communists" and of interfering too much with the prerogatives of the settlers in Algeria.[7] Indeed, he was singled out as a target by the extreme elements of the French Right, as he was to discover after the uncovering of the plot perpetrated largely to discredit the Mendès-France government and known at the time as the *affaire des fuites*. In the spring of 1954, it became generally known that very secret proceedings of the highly exclusive National Defense Committee, attended by only ministers and very high brass, were being leaked. Rumor had it that a minister, or ministers, was betraying secrets to the Communist party. To make a long story short, it was finally discovered that while the leaks were real—they were the work of two civil servants—the alleged reports of the political bureau of the Communist party, into which they had been woven, were forgeries, the handiwork of a dubious double agent that was peddled by a police inspector of extreme right-wing views for the obvious purpose of compromising the government and, in particular, its minister of the interior. Mitterrand, while emerging this time unscathed, learned how low the Right was ready to stoop if its real interests were at stake.

How ambitious Mitterrand really is cannot be adequately assessed by lumping his whole career together. Eleven times minister during the life span of the Fourth Republic, which lasted only twelve years, even allowing for a quick turnover in governments, this shows a healthy appetite for office and honors. Taking this as an indictment, the accused pleads that only once did he serve in an (openly?) right-wing government, that of René Laniel, and that he resigned from it within three months. To which it may be retorted that he did not resign, as did Mendès-France, from the hardly more honorable "Socialist" government of Mollet, the man of Suez and of torture in Algeria. This is the reason why he gained at the time the reputation of a "Florentine"—the expression was first used by the novelist François Mauriac—keener on Machiavellian politics than on principle, an expert on the electoral map, on party permutations rather than a man of historical vision. (It became a habit on the Left to contrast Mitterrand, the clever politician, with Mendès-France, the moralist. To which pragmatists replied that the latter became an admired spectator; the former, the main actor in the political drama.)

To understand how the master tactician became a strategist, how a man who, until the age of forty-one, could barely stand to stay out of office spent the next twenty-three years in the wilderness, how the clever politician opted for a historical role rather than immediate honors, we must stop for a moment to look at a crucial point in Mitterrand's career—General de Gaulle's return to power in 1958. Driven by the need to face European competition within the projected Common Market, the tottering Fourth Republic could no longer afford to carry the burden of a colonial war. The French settlers in Algeria, the colonial lobby, and the military commanders of a North African army 500,000 strong refused to listen to historical logic. On May 13, 1958, they rebelled and took over in Algiers. Parliament in Paris, refusing to yield, appointed the Christian Democrat Pierre Pflimlin as prime minister. This is when General de Gaulle came openly onto the stage, aiding and abetting the rebels, but also offering to parliament and country his services as national savior. It did not take long for Pflimlin, President René Coty, Mollet, and company to surrender.

Writing with hindsight but also, by then, from a much

more radical viewpoint, Mitterrand argues that such a conclusion was in a sense inevitable because both sides, Paris and Algiers, were frightened above all of "awakening popular passion."[8] The bourgeois rulers in 1940 had feared Hitler less than the popular front and, in 1870, the Prussians less than the Commune of Paris. Now the Establishment of the Fourth Republic "felt closer to those who were assaulting it than to those who could save it—the people in arms." In other words, fundamentally, there was a possible answer, there was a parliamentary majority—including Communists and Socialists—to carry it out, and there was a man ready to take the necessary lead—Mitterrand himself.

Let us elaborate. First, the mood. Admittedly, Mitterrand is commenting over ten years later and shortly before joining the Socialist party. Yet the way in which he describes a break with his milieu at just that time sounds quite convincing:

> In Jarnac, in Nevers, in my Paris street, in the path of my *stalag* I had met them. In a certain way I was one of them. Because we had the same origins, because we pronounced the same words in the same fashion, because we had received the same rudiments of culture, they treated me as an accomplice and an associate. Until the day I knew we had nothing to say to one another.

The Fourth Republic was collapsing, and Mitterrand goes on to say how shocked he was at first over the popular indifference to its fate. On reflection, he does not blame the people, but the "mandarins who did wear out the words and empty things of their substance." Then comes the self-criticism: "And I had taken part in this enterprise, respected its habits, did not shout loud enough to upset the ceremonial." This was now over. He would not behave as had so many others, not caring about the number of the Republic, provided it preserved "the structures, privileges and profits." Thereupon, he went back to the National Assembly to vote against General de Gaulle: "To preserve honor was the only way to await in peace with myself the end of contradictions."

Second, the facts. Mitterrand refers twice to conversations he had with Coty, the president of the Fourth Republic, toward the end of this trial of strength between Paris and Algiers. In the first reference, Coty told him that he often thought of calling him to solve the crisis, but decided not to "because there would have been incidents in Algiers."[9] In the second, more recent, reference, he was more specific: "René Coty sent for me and asked: 'Would you accept Communist votes?' I answered: 'I certainly would, and, if that were not enough, I would solicit them.' He then told me: 'It's impossible.' "[10]

All the pieces of the puzzle now fell into place. The young politician saw the opportunity of finally getting to the top: "to say that at last I was going to be prime minister," he apparently exclaimed to a friend.[11] But this was a different kind of battle. In the heat of it, he had to stop dillydallying; he had to choose his side. At the same time, he perceived a historical chance, the prospect for higher ambition and the strategy likely to take him to the very top. The Left could win in France if it were united—that is to say, if it counted the Communists in its ranks. Its victory, however, would be accepted, tolerated internationally as well as domestically, only if the non-Communist Left gained the upper hand in this coalition and the Communists agreed to be junior partners. Mitterrand may not have known at the time that to achieve his purpose he would have to be converted to socialism, but he was convinced that the game was worth playing and that he was the man to win it. A new life was beginning at just over forty.

Conversions are never entirely sudden and complete. Indeed, the first major event in his life after this turning point seems to fit a clumsy politician rather than a statesman, although, on second thought, it might have broken a lesser man. On October 16, 1959, the French learned that on the preceding night Mitterrand had miraculously survived an attempt on his life. Driving home—he was already living in the Latin Quarter, although closer to the Luxembourg Gardens—he had to jump out of his car, which was machine-gunned, climb over the fence of the neighboring Observatory Gardens, hide behind bushes, and what not. Even in the context of the Algerian war, with rumors of commandos of killers crossing the Spanish frontier, such an attack on one of the leaders of the opposition hit the headlines. Exactly a week later, it turned out that it might have been a hoax. A certain Robert Pesquet, a shady character and former Poujadist deputy now connected with extremists of "French Algeria," argued with some proof that he had arranged the whole affair with the victim. Mitterrand had to admit that the man had come to see him, alleging that he had been sent to shoot him. The attempt was real for Mitterrand, who concealed the man's name from the police to protect him, wrongly assuming that he had warned him because of a case of conscience.

Whatever the details, it is difficult to question Mitterrand's version globally. After all, he obviously did not take such risks simply to hog the limelight. Besides, he did not come off too well from the whole business, the clever politician being bamboozled like a political babe. I am less worried by the veracity of his version than by the hints and insinuations produced by the accused when in a tight corner. He suddenly remembered, in public, that less than three years earlier, when he had been minister of justice, he had treated much better a Gaullist politician—who had been mixed up in a much bloodier *affaire,* the so-called bazooka case[12]—than he, Mitterrand, was now being treated by the government. And the man in question was none other than Michel Debré, presently de Gaulle's prime minister. There was here more than a hint that the mighty deserve special treatment, that there are rules for them and if they are broken . . . There was a sneaking suspicion that the former minister of the interior and of justice was not a man to leave without files and ammunition. Maybe one had to behave like that, since the other side clearly spared no effort to crush its potential enemy. Mitterrand, having concealed the truth, was sued for "contempt of the judiciary," although the case was never brought to court. It was kept hanging over his head as a threatening weapon. It took time and his proverbial resilience for the discredited Mitterrand to bounce back to center stage.

Vae victis. In the November 1958 election, following General de Gaulle's triumphant return to power, most of the leaders who had dared to oppose him were defeated. Mitterrand was one of them, although he fared better than most, staying only five months out of parliament. Rejected from the lower and, in France, crucial chamber—the National Assembly—he climbed into the upper house—the Senate. Senators in France are chosen indirectly by regional and local councilors elected earlier and, in this case, unaffected by the Gaullist wave. Mitterrand had by then a sufficiently strong position locally to be picked as senator for the Nièvre. More significant than the British House of Lords, the French Senate cannot be compared in political importance with its American namesake. It was useful for Mitterrand as a temporary platform. In the general election of 1962, the Gaullist wave having partly receded, he managed to be reelected as deputy. Back in the National Assembly, he became the most eloquent censor of the new regime and was well placed to wait for the main chance.

With the decision to elect the president of the Republic by universal suffrage, the presidential poll of 1965 dominated French politics and put the non-Communist Left in a dilemma: To challenge General de Gaulle, should it look to the Right or to the Communists for support? Conditioned reflexes drove it to revive the third force between Communism and Gaullism. Launched like a soap powder, first as Monsieur X,[13] the Socialist Gaston Defferre tried to reach an alliance with the Christian Democrats. This coalition, most unlikely to do well, never came off. The negotiations dragged on, and when they finally collapsed, there was little time left before the start of the campaign. Mitterrand then moved to fill the vacuum. Although his strategy was quite contrary—he himself was in favor of an alliance with the Communists—he had played fair with Defferre and was now able to ask for similar favors. Because nobody was particularly eager to enter what looked like a dubious battle, Socialists, Communists, and Radicals—in that order—accepted Mitterrand as a candidate. He, as we saw, did better than expected. He not only emerged from this confrontation as the mythical symbol of left-wing unity, a most precious capital for his political future, but also appeared as the leader of a coalition likely to win—if not tomorrow, then the day after. The 45 percent of the vote gathered in single combat with the legendary hero suggested that in the ordinary parliamentary contest the Left would do better. And it did. Two years later, in March 1967, the right-wing rulers clung to power by only the skin of their teeth or, to be more precise, thanks to the modern version of the nineteenth-century rotten boroughs, the overseas constituencies, where miracles are more easily staged. In electoral terms, Mitterrand now seemed to be the inevitable future winner.

Only the terms refused to be electoral. Mitterrand was no more bewildered by the 1968 upheaval than were other traditional politicians. I remember seeing him in the Latin Quarter on one of those bloody Fridays when barricades were burning, surrounded by his lieutenants, totally ignored by the students, and peering through the gas-infested atmosphere like a man from another planet. However alien the movement, he attempted to use it to his own political advantage. In his press conference on May 28, he argued that should General de Gaulle be defeated or resign, there would be no void: The Left would fill it with Mendès-France as provisional prime minister and himself as potential president. While not particularly impressive as crystal gazing, this was in no way illegal talk, and the Gaullists, who ten years earlier had shown utter contempt for the laws of the Fourth Republic, looked phoney and ridiculous in their new posture as indignant censors and righteous upholders of legality.

If Mitterand was subsequently treated as an outcast, even in his own camp, it was that the prospects had altered, at least temporarily. With victory no longer within immediate grasp, those on the Left who had accepted him reluctantly, as a lesser evil, were now ready to push him aside. The May events, showing that in the factories and in the streets the Communists weighed much more than the Socialists, undermined a vital element in his political premise. On this occasion, however, recovery was not too difficult an exercise. By putting up two candidates in the battle for de Gaulle's succession and losing disastrously, the non-Communist Left demonstrated that divided it would not balance the Communist party, let alone challenge the Right. In a negative way, it showed that there was no alternative to Mitterrand's strategy. It was now up to him to resume the struggle in a more comfortable position, as the official leader of the Socialist Left.

June 1971 is a date to remember. The founding congress of the new Socialist party, a regeneration of the old SFIO, at Epinay, just outside Paris, was both a historical and a very strange occasion. Puzzling was the alliance among, on the one hand, Pierre Mauroy, the mayor of Lille (and future prime minister), Gaston Defferre, the mayor of Marseilles (and future minister of the interior)—two bosses of social-democratic bastions traditionally opposed to an alliance with the Communists—and, on the other, Jean-Pierre Chevènement (future minister of industry), leader of CERES, a leftish pressure group that favored self-management and a popular front with the Communists. Puzzling, too, was the climb of Mitterrand, at the head of this coalition, to the post of first secretary, the newcomer getting the leader's insignia and his membership card almost at the same time. Stranger still, however, for anyone who had followed his career throughout, was the language of class struggle adopted by the convert.

"Reform or revolution? . . . yes revolution" could be dismissed as congressional rhetoric, if the orator were not driving the point home:

> Violent or peaceful, the revolution is first of all a break. He who does not accept the break . . . with the established order . . . with the capitalist society, he, I say, cannot be a member of the Socialist party. . . . There is no, there will never be a socialist society without the collective

appropriation of the great means of production, exchange, and research.

And to dispel remaining doubts, this peroration on the degrading power of capital: "money that buys, money that crushes, money that kills, money that ruins, and money that rots the very conscience of the people."¹⁴

"M. Mitterrand has learned to speak socialist," commented, with the irony of a defeated rival and the knowledge of an expert, Guy Mollet, who was capable of justifying colonial expeditions in socialist language. The zeal of the convert, the atmosphere of a congress, and the general mood, sharply radicalized by the echoes of 1968—all this must naturally be taken into account. It is not enough to dismiss altogether this radical strand, which for the next ten years would return as a refrain in Mitterrand's pronouncements and publications. Just a glimpse at extracts from his diaries will reveal a series of references to the need "profoundly to modify the very structure of our society." The answers to the questions about the man's sincerity and pragmatism, about the depth and versatility of his political convictions, must be left to the conclusion of this book. Here suffice it to say that this strand, although not exclusive, was significant. On the eve of the presidential contest, in 1974, Mitterrand jotted in his diary that Giscard, if elected, might accomplish fine things, but only he could "change the course of things and therefore the life of my contemporaries."¹⁵ He was relying on the Socialist transformation to leave his mark.

Mitterrand's defeat in 1974, while psychologically important, was not politically crucial. Obviously, it was maddening to fail by a fraction of 1 percent, and Mitterrand, nearly fifty-eight years old, mused sadly over his future. In a moment of depression, he told his closest assistants that although victory would be theirs next time, he would no longer be there to lead. However serious or moody that forecast, Mitterrand did not doubt ultimate success, because defeat in no way affected his basic premise. The political crisis of 1977 and 1978, on the contrary, did. With the Communists openly rejecting the idea of being junior partners in the left-wing coalition, the strategy that had inspired his conduct for two decades was torn apart. The Socialist party was being urged to cast aside the strategy and drop its architect by the same token. This was the moment when Mitterrand showed that tactics can be raised to a fine art. He categorically refused to abandon the search for a united Left. He thus gave the impression of a man of principle, of deep conviction. He also revealed himself to be a political champion with an eye on the main chance. He chose the only road that could lead to victory and now got his reward.

Sweet is revenge, say the poets in most languages, and Mitterrand was undoubtedly savoring his. Revenge over fate, which for so long had seemed to defy him; revenge over his arrogant, pseudoaristocratic rival, who had defeated him seven years earlier; revenge over many "comrades" who barely a year earlier had wanted to bury him politically, pushing him like a relic into the Panthéon. All this was most enjoyable, yet trivial with the moment of truth approaching. French lore has it, at least since the creation of Rastignac by Honoré de Balzac, that ambitious young provincials come to Paris, climb to the Sacré-Coeur in Montmartre, and from there issue a challenge to the capital. But Mitterrand, in his sixty-fifth year, was no longer a youngster, and in his ambitious project of preparing the ground for an admittedly vague form of democratic socialism, he was challenging not just Paris and France, but also the whole of Europe and the world at large. He had expressed his contempt for those who, having used up "all their energy on a career have nothing left for history."[16] It was his turn, having climbed to the top, to show whether he had enough in him to leave a mark.

II

The Road to Surrender

5

The Fall from Grace

> The institutions were not made with me in mind, but they were made for me.
>
> François Mitterrand
> in an interview given to
> *Le Monde,* July 2, 1981

It all began beautifully with a carefully staged show. The Left was not satisfied with the routine. On May 21, 1981, the usual takeover ceremonies at the Elysée, the presidential palace, and the ride of the new head of state up the Champs-Elysées were only the beginning. In the afternoon, after a stop at the city hall, François Mitterrand crossed the Seine on his way to the Latin Quarter, the academic heart of the capital. In an open car, he drove up the Boul'Mich, the main avenue and witness of yesterday's student battles, stopped in front of the Luxembourg Gardens, and from there walked on, a rose in hand, surrounded by friends French and foreign. Paris was for a while Europe's Socialist capital, with former prime ministers—including Willy Brandt of West Germany, Olaf Palme of Sweden, and Mario Soares of Portugal—mixing with premiers of tomorrow—including Felipe Gonzales of Spain and Andreas Papandreou of Greece—all gathered for the occasion. The French president also insisted on having his literary friends among his personal guests. Arthur Miller and William Styron had come from the United States; Julio Cortázar of Argentina, Carlos Fuentes of Mexico, and Gabriel García Márquez of Colombia represented Latin America. And behind all these distinguished figures marched the people of Paris.

In front of the Panthéon, where the French bury their illustrious dead, the Orchestre de Paris, conducted by Daniel Barenboim, played the Berlioz version of the "Marseillaise" and Beethoven's Ninth Symphony, with its choral "Ode to Joy." The crowd had to stay outside, and it was the turn of television viewers to watch the sophisticated performance as the lonely president laid roses on symbolically selected tombs: that of Victor Schoelcher—even the French had to consult their encyclopedias—who proclaimed the end of slavery in 1848; that of Jean Jaurès, the great Socialist orator and fighter for peace who was murdered by a warmongering fanatic on the very eve of the First World War; and that of Jean Moulin, the underground leader who was tortured to death by the Nazis. Libertarian legislation, democratic socialism, and the Resistance were thus cleverly woven into one. At the end of the ceremony, Placido Domingo sang France's Revolution-inspired national anthem, and as the president left the house of the illustrious dead, the crowd, breaking barriers, got out of hand. Yet it was all good humored. For once, the confrontation between students and police in the shadow of the Sorbonne was friendly. The lay Mass devised by Jack Lang[1]—yesterday, professor of law and man of theater; tomorrow, minister of culture—had been a tremendous success, and when it was well over the young people of Paris went on dancing in the rain.

For Mitterrand, it was back to business at once. On the very morrow of the inauguration, he had to form his government and announce a new election to the National Assembly. The latter he did not quite have to. Some Socialists—Michel Rocard, for instance—were in favor of postponing the poll, possibly on the unavowed ground that coexistence between a Socialist president and a conservative parliament would inevitably lead to compromise and moderation. The French president, as we saw, unlike the American, has the power to dissolve the crucial lower house, the National Assembly. Mitterrand, who had told friends that euphoria—the "state of grace," as he had called it—is by definition transient, was determined to make use of it. Freshly elected, the president was in an excellent posture to ask the French people to give him the parliamentary means to carry out his policies. In an excellent position, too, to exploit the unhealed wounds of the presidential poll—the divisions between the followers of Valéry Giscard d'Estaing and those of his rival, Jacques Chirac.

Pierre Mauroy, the prime minister he chose to lead the electoral campaign, was a proof that Mitterrand was capable of forgiving. The fifty-two-year-old mayor of Lille had taken Rocard's side in the struggle, two years earlier, over who would be leader of the Socialist party and its presidential candidate. But he had since made amends and was back in favor. Tall, big, and cheery, Mauroy looked reassuring to French and foreigners alike. The industrial north is one of the regions where, with proletarian connections, French Socialists have something in common with the social-democratic tradition of northern Europe. A product of this region, the pragmatic Mauroy could in no way be painted as a scourge of capitalism. At the same time, he was recognized on the Left as somebody belonging to the family. Son of a teacher and once a lecturer in a technical college, he had spent most of his life within the Socialist party. For long a leader of its youth movement, he had climbed the ladder under Guy Mollet, prepared the way for Mitterrand's victory, and served then as his second in command. In France's division into two nations, Mauroy, however moderate, obviously stood on the left side of the fence.

For foreigners likely to see red, there were other guarantees in the new government. The Foreign Office, or to give it its new title, the Ministry of External Affairs, was entrusted to an expert recalled from Brussels. At sixty-one, Claude Cheysson had a long career behind him. Although his preoccupation with the Third World may have perturbed some people in Washington, he had been delegated to the Executive Commission of the Common Market by Mitterrand's conservative predecessor, and, in any case, European commissioners are not exactly the stuff of which revolutionaries are made. More important still, the crucial Ministry of Economy and Finance was given to one of the most moderate of Socialist leaders, the fifty-five-year-old Jacques Delors, who in 1980 had had the honesty to resign from the Socialist party executive because he thought its program was too radical. His unconcealed ambition was to modernize and reform to some extent capitalist society, only this and nothing more. His previous day of glory had actually been on the other side. When, in 1969, President Pompidou had decided that social peace required a sort of New Society government, Delors had been the chief adviser on social policy of Jacques Chaban-Delmas, the prime minister. With such a man now in charge of economic policy, the men of property, French or foreign, had no need to panic. Add to it that planning in the new government was put in the hands of Rocard, who in the inner Socialist controversy had insisted on the primacy of the market. Mitterrand's forgiving nature should not be exaggerated: If Rocard was given that ministry, then planning had not much future.

Reassurances for foreigners were on the agenda because the forthcoming election raised the issue of Communist participation in the coalition and, hence, in the government. In the prevailing French electoral system, as we have seen, in the first ballot voters can express their preferences; in the second alliances are struck to defeat opponents. In the bulk of the 491 constituencies, duels were usually fought between the respective champions of Left and Right. Such streamlining was possible thanks to alliances normally reached on a national scale. In the three previous years, Communists had played hard to get. In the new circumstances, they were eager to please in order to be allowed to climb on the bandwagon. The Mauroy government's raising, as promised, the minimum wage, old-age pensions, and family allowances set the stage for a snap election.

Its result exceeded even Mitterrand's expectations. His presidential success was turned into a parliamentary landslide. In the previous election to the National Assembly, in 1978, the Left had had a small edge in the total vote on the first ballot; a bad transfer of votes between Socialists and Communists, combined with a slight conservative bias in the drafting of constituencies, meant that, after the second ballot, the Right had 290 deputies to the Left's 201. This time, many more people stayed at home. Thus the Left did not really gain votes; only the Right lost them on a mass scale. As a result, in the first ballot, on Sunday June 14, the Left reached an unprecedented share of the vote—55.2 percent of the total—with all shades of the Right reduced to 43.2 percent (the remaining percentage was accounted for by people running on an ecological platform). The following Sunday, in the second ballot, this conservative disaster was translated into seats. The followers of Chirac, Giscard, and all their allies had only 157 deputies to the Left's 334. The real triumph, however, belonged to the Socialists. Together with their junior allies, they captured about 38 percent of the votes cast in the first ballot. In the new chamber they could rely on some 290 deputies (270 of whom were members of the Socialist party). The Communist party had 44 deputies.

For the first time in its history, the Socialist party had a comfortable majority in the National Assembly on its own, without Communist support. Yet in the reshuffle of the Mauroy government after the electoral victory, the Communists were brought in. They were given one nominally senior post, that of minister of state in charge of transport; since Georges Marchais stayed out, it was taken up by forty-seven-year-old Charles Fiterman, who had started work as an electrician at the Schneider engineering plant in St. Etienne, but had since climbed all the rungs of the ladder in the party apparatus. The other three portfolios granted to the Communist party were the Ministries of Professional Training, of Health, and of the Civil Service.[2]

Why did the Socialists bring the Communists in when arithmetically they were dispensable? Why not? The reply of the March Hare from *Alice in Wonderland* is more relevant than ever. The price paid was low. With four ministries out of forty-four, and none of them key ones, the Communists would carry little weight and yet would be bound by official decisions. Although the parliamentary election confirmed its decline—its share of the vote dropping from 20.6 percent in 1978 to 16.2 percent—the Communist party preserved strong roots in the working class and control of the CGT, the biggest trade union. As a guarantee against trouble on the labor front, the inclusion of the Communists in the government was a very small fee. For Mitterrand, the move was not only consistent with his policy, but also very convenient. The real question is why the Communists accepted this solution of responsibility without power; they could have joined the coalition without entering the government and thus vote on only the measures they considered progressive and popular. The answer is, probably, that bruised and battered by their contradictory acrobatics, they assumed, wrongly, that a cure in government would restore them to fitness.

However innocuous it looked from Paris, the admission of the Communists into the French government was hailed in Washington as a symptom of the "Red peril." As Vice President George Bush entered the Hotel Matignon, the residence of the French prime minister, the ushers had to perform one of the classical acts of theatrical farce so that he would not bump straight into Fiterman, the peril personified. Bush allegedly told Mauroy that the Americans were worried not so much about the French prospect as about the precedent itself, with Italy particularly in mind.[3] The State Department was not so diplomatic, since the very same day the French could read on their teleprinters that "the style and contents of relations between the two countries would be affected." And there was plenty to come. Jacques Attali, the intellectual prodigy acting as Mitterrand's personal assistant at the Elysée, was apparently asked by one of Reagan's advisers to explain to him "the real difference between your economic policy and that of the Soviet Union."[4] Delors claims that as soon as the Left got into office, France was put in quarantine. Its new government was the object of systematic suspicion: "When they took part in international meetings Socialist ministers were looked upon as if they had arrived from another planet, a red flag flying in their hand."⁵

The belief that the Socialist dog would be wagged by its Communist tail seems too absurd to be genuine. It was, however, possible to imagine that without prodding from the Communist party, the new government might, on its own, have the will and the means to introduce policies likely to disturb or even to endanger the Western capitalist Establishment. The narrative so far, emphasizing the Socialist triumph, might have given the impression that the Mitterrand government's freedom of action was boundless. It is, therefore, necessary to glance back at the seamy side of the story to recognize that constraints and limitations were there from the very start.

On May 11, Mitterand's election and the victory of the Left were greeted with perfect symbolism. The Bourse, the Paris stock exchange, was literally paralyzed: With plenty of sellers and no buyers, there could be no quotation. On the currency market, the French franc hit the bottom: It dropped to its floor level within the European Monetary System (EMS), which it could cross only by leaving the system or by an official devaluation.⁶ The fall of the shares was sudden, as though the Bourse had been assuming Giscard's victory. Speculation against the franc, on the contrary, had been rife in the run up to the election. It is estimated that between the beginning of February and May 10, about $5 billion left the country. During the interregnum, Raymond Barre, the outgoing premier, refused to go beyond the duties of a caretaker, and the flight continued: Between May 11 and 15, another $3 billion vanished; in the week beginning May 18, daily losses climbed from $500 million to $1 billion and reached probably $1.5 billion on May 21, the day of the inauguration. Then, as they were riding up the Champs-

Elysées in an open car, the new prime minister tried to raise the urgent issue with the president and was told that this was not the moment. In between ceremonies, however, Mauroy obtained Mitterand's approval for a policy of defending the franc and refusing a devaluation.[7]

Renaud de La Genière, the governor of the Bank of France, was pleasantly surprised by the Left's decision not to devalue. Drastic measures were taken at once to bolster the franc. The bank rate was raised to 18 percent, and strict exchange controls were introduced, with allowances for French tourists and remittances by immigrants working in France the only exceptions to the rule. It nevertheless took three more days, a visit by Chancellor Helmut Schmidt, and his expression of West German backing for the French currency for the speculation to abate, at least for the time being. The left-wing upstarts had been warned.

Retrospectively, the decision to cling stubbornly to the inherited international value of the franc has been taken as a sign of political abdication, the first step on the road to surrender. This may be going too far, although the move was clearly political. On purely classical financial grounds, the differential in the rates of inflation between France and Germany, its chief partner, estimated at about 14 percent since the last currency reshuffle, justified a substantial devaluation of the franc, and a new government always has the opportunity to put the blame on its predecessor. If its partners in the EMS had refused to agree to the terms proposed, France could simply have left it for a time. There was nothing terribly "subversive" in such an exit. Giscard, a respectable conservative, took the franc out of the so-called snake of European currencies for a period, and Britain, even under Margaret Thatcher, did not enter the system. The attitude toward devaluation is not, on its own, a criterion of radicalism. Should a Socialist government in France, or elsewhere, take really radical anticapitalist measures, its currency—whether fixed or floating, within or without a specific monetary system—would come under international counterattack. Indeed, floating the franc might have helped a moderately reformist government, giving it slightly more scope for a Keynesian reflation. The new French regime did not take this option to float in order to show its orthodoxy, its respect for bourgeois niceties, and the international rules of the game. Only in this sense can the nondevaluation be described as the beginning of resignation.

And it can be so described only with hindsight. At the time, in the first months of its existence, far from betraying its pledges, the Socialist government was literally fulfilling them. To the surprise of some people, assuming that the Socialists forget their promises and principles on the very day they enter office, President Mitterrand was proclaiming time and again that his government would carry out "nothing but its program, but the whole of its program." It was to do so in the full majesty of the law, respecting all existing formalities, all regulations. All the shareholders in the industries to be nationalized were to be given ample compensation. The opposition was to be allowed to slow down the parliamentary process through filibuster, as it had never been allowed before. The recommendations of constitutional watchdogs were to be obeyed religiously. Although inevitably hampered by all these handicaps, the newcomers in the ministerial offices were busy drafting a vast legislative program designed to translate the bulk of Mitterrand's 110 proposals into practice.

Rather than study their fate one by one, it is preferable to regroup these measures under four broad headings. First, the Left had promised to improve the lot, if not of the downtrodden and humiliated, at least of the unprivileged part of French society. Ever since the Second World War, people working in France have had a legally guaranteed minimum wage, which must at least keep up with the cost of living. In June 1981, this Salaire Minimum Interprofessionel de Croissance (SMIC) was raised by 10 percent, instead of the 3 percent required, because of inflation, boosting the wages of some 1.5 million working people. Simultaneously, around 2 million people were affected by a 20 percent increase in old-age pensions and about 3 million by a 30 percent rise in family allowance. Add to it, after the elections, the commissioning of 61,000 jobs in the public service, part of an original plan to create some 210,000 jobs and thus begin to deal with the major task of all, the struggle against unemployment. Various measures to reduce working time, desirable in themselves, must also be seen as helpful in this connection.

In this field, the French Left has a tradition. It began in 1936 when the popular-front government broke utterly new ground, introducing two weeks of paid vacation for everybody. Films of that period show special trains taking the workers on their first journey to the sea. Now it was the turn of Mitterrand to extend this vacation from four weeks to five. France is in this respect a pace setter: All working people are entitled, by law, to five weeks of paid vacation every year. Another connected reform, the right to a pension at age sixty, proved more complicated. Life expectation varying with job and income, as long as retirement was fixed at age sixty-five, few industrial workers enjoyed their pension for long. Thus the Socialist plan to give the choice of retirement at age sixty, with a full pension after thirty-seven and a half years of work, was welcome. Trouble began with the level of that pension. The previous government, trying to find room for the unemployed, subsidized a scheme of early retirement. In some cases, its terms were more favorable than those of the new plan open to all. The unions protested, and the terms had to be renegotiated. The reform is popular—and any administration will have difficulty going back on it—but it was not greeted with the expected enthusiasm. The third reform in this field, the proposal to reduce the work week within five years from forty to thirty-five hours, began timidly and then totally flopped. In 1982, the week was shortened by one hour, amid disputes whether everybody should benefit from it without loss of salary, and then the whole project vanished, together with many left-wing illusions.

The retreat of the Socialists when faced with economic reality will be dealt with in Chapter 6. Here it is enough to say that the social bill, although substantial, was not extravagant. In financial terms, the package, designed at the same time to spur output through higher consumption, was not comparatively bigger than the one introduced by Chirac's government after Giscard's election in 1974. This having been said, the circumstances were quite different. With all the Western countries deflating and the welfare state under attack, the French Left dared to go against the trend. In this early period, the Socialists were sticking to the commitments they had made to their own electorate.

The second series of measures was connected with personal freedoms and their general legal framework. They are loosely linked with the man who can be described as the most successful or the most hated member of the government. When he joined it as the minister of justice at the age of fifty-three, Robert Badinter had a brilliant career as a lawyer behind him. He was not a professional politician. On issues like nationalization and international monetary policy, he was rather moderate. He is not so much a socialist as a progressive liberal in the American sense of the term, a staunch defender of civil rights and a man of principle. Having passionately opposed the death penalty as a counsel, he remained true to himself on getting into office; by October 1981, the death penalty had been abolished, France incidentally being one of the last countries of Western Europe to get rid of this barbarian vestige. He also presided at once over the suppression of all forms of exceptional jurisdiction, such as the Court of State Security.[8] If you add on to it initiatives taken in other ministries—the regular papers granted to previously illegal immigrants,[9] the instructions presumably given to the police not to be guided by the color of the skin in their identity checks, and so on—for a brief spell you had in Paris the feeling that ballot papers may have the capacity for altering a country's general climate.

It did not last too long. As soon as Badinter began to dismantle the handiwork of Alain Peyrefitte, his conservative predecessor—especially the law ironically called "liberty and security," in which the former was sacrificed in vain on the altar of the latter[10]—he ran into trouble. Although he carried the day, the resistance he had met was a portent of things to come. Son of deported Jewish parents, Badinter was not only a perfect target for Jean-Marie Le Pen's National Front, but also the obvious scapegoat for the growing law-and-order brigade, which, oblivious to facts and figures, blamed this "lover of criminals" for the insecurity and the fears in overcrowded towns, an atmosphere getting worse with the economic crisis. Some Socialist leaders would have loved to get rid of such an inconvenient symbol. Already at this stage, it must be said that Mitterrand, rising in this respect above the immediate electoral calculations, did back his minister of justice throughout. Whether Badinter was the honor of Socialist conscience or merely an alibi must be left to the final assessment.

The third series of measures was concerned with the structure in which political and social activity is carried out, and pride of place was given to the proposed transfer of some power from the capital to the provinces. Indeed, the ministry entrusted to Gaston Defferre, the veteran Socialist mayor of Marseilles, was renamed the Ministry of the Interior *and* Decentralization. From the time of Louis XIV and the absolute monarchy, through the Jacobins of the Revolution and Napoleon, to the weak bourgeoisie of the nineteenth century, unable to rule on its own, France required a stronger state than most Western countries, and the "Jacobin" Left had the reputation of being the main upholder of this centralizing tradition. Mitterrand and his colleagues, while widening the public sector, were determined to prove that their conception of socialism was not to be confused with the extension of central state power. They introduced a good deal of legislation to strengthen local autonomy as well as to transfer decision-making power and financial resources from Paris, notably to the twenty-three regional assemblies elected through direct suffrage. As midterm elections swung against the Socialists and they lost many of their local positions, their zeal for decentralization inevitably cooled. They nevertheless carried out the proposed reform. Again, it involved no upheaval, no major change in the way France is governed. But the Socialists did stop and even slightly reverse a secular trend, and that is not negligible.

Other measures designed in principle to alter the framework did not go very far. The Socialists were supposed to relax the strangle hold of financial interests on the press. They had also promised to include most private schools in a vast reformed and unified system of public education.

They hesitated so much before moving forward and then dragged their feet in search of a compromise for so long that the climate altered in the meantime. They themselves contributed to this conversion, to the new belief that private is beautiful and the market a synonym for freedom. Their glorious projects then just collapsed. These important defeats, however, are not the subject of this chapter.

Finally, we come to the reforms that aroused the greatest passion, linked as they were with property and social relations at the work place—the nationalization projects that led Reagan's adviser to equate the French economy with the Soviet, and the so-called Auroux laws, from the name of the minister in charge, designed to give new rights to the working people in their factories.

The role played by the public sector in a country like France is best approached through an anecdote. When in 1982 industrial and financial firms were finally nationalized amid much noise in the media, an elderly woman went to the local branch of one of the big French banks to close her account. The manager, while filling out the forms, asked her politely for the reason. "I can't entrust my savings to the state," replied the woman, with the righteousness of somebody stating the obvious. "But madam," replied the manager, "you have been doing it ever since the war." The anecdote stresses the fact that it was after the Second World War that General de Gaulle, inspired by the Resistance movement, carried out the great transformation and that the French had since become used to living in a landscape of which public property was an intrinsic feature. The bulk of schools and hospitals, the post office and telecommunications were state owned. So were the big banks, the railways, the gas and electricity utilities—actually quite a strong nuclear lobby—as well as the Renault car works, nationalized because its owner had collaborated with the Germans. In carrying out their program of nationalization, the Socialists were not pioneers. They were expanding frontiers of an already established public sector.

The whole operation took an unexpectedly long time, although the government embarked on it at once. Mauroy, with the president's backing, had first to persuade several ministers and most of the economic advisers that the state must take over the selected enterprises wholly and not simply gain a majority shareholding.[11] Once drafted, the nationalization bill was presented to parliament on October 13, and there met unprecedented opposition: 118 hours of obstruction rather than debate were required in the National Assembly. When a Socialist deputy, Michel Berson, accused the Right of fighting so hard because it was linked by family ties with the big shareholders, he provoked pandemonium. Vehement protestations were superfluous in a way, since if there were no family ties, the Right's passionate defense of class interests was all the more revealing. By December 18, the bill was finally through both houses (the Senate in France can only delay, not block, legislation). It still had to run the gauntlet of the Constitutional Council. At the time, none of the nine members of the council had been appointed by the Left.[12] The council did not follow the advice of some sections of the Right to throw the law out altogether on the grounds that it clashed with the Declaration of the Rights of Man of 1789, including property among those "natural and imprescriptible" rights. It did, however, question the pretty generous compensation, thus forcing the government to rewrite the bill and rush it again through parliament and also increasing the already heavy burden of the operation by over 20 percent.[13] The bill became law in February 1982, and only then could the Socialists appoint new heads of the nationalized firms, and start reshaping what Mitterrand had hailed as the "strike force" of the French economy. The delay was a heavy tribute to pay for respectability.

The extension of the public sector was important, particularly in industry. The state acquired 100 percent of the stock of five major industrial conglomerates: the Compagnie Générale d'Electricité (CGE), with 180,000 employees, principally in electronics, engineering, and telecommunications; Thomson-Brandt, with 129,000 employees, mainly in electronics for consumers and military equipment; St. Gobain, with 136,000 employees in glass, engineering, and electronics; Rhône-Poulenc, with 89,000 employees in chemicals and textiles; and Pechiney, Ugine and Kuhlman (PUK), with 86,000 employees in aluminum and chemicals. In November 1981, the two big steel corporations Usinor and Sacilor had been taken over by just turning past debts into property rights, an easy exercise because the amount of public money poured down that private drain had been a notorious scandal, "the biggest since Panama."[14] Then the government negotiated a majority shareholding in two firms working mainly for the Ministry of Defense: the Dassault aircraft company and Matra, which sold its stake in the media. Although it was slightly more complicated to nationalize foreign-owned firms, the French state bought 50 percent of Roussel-Uclaf from Hoechst, the German pharmaceutical giant; reduced the American share from 47 to 20 percent in CII Honeywell Bull, the computer firm; and purchased from International Telephone and Telegraph its French subsidiaries, the Compagnie Générale de Construction Téléphonique (CGCT) and Laboratoire Central de Télécommunications (LCT).

In general banking, the advance was less impressive because the big joint-stock banks had been nationalized after the Second World War. The spectacular move was the takeover of Paribas and Suez—the Compagnie Financière de Paris et des Pays Bas and that of Suez—two merchant banks that were symbols of French capitalism. Otherwise, the nationalization of thirty-six smaller banks brought the share of the state from 64 to 74 percent of all deposits and from 56 to 69 percent its share of credit advances.

In industry, in calculations based on data for 1980, the part of the state climbed from 17 to 29 percent of total sales, from 11 to 22 percent of employment, and from 44 to 52 percent of total investment.[15] The difference would have been bigger if the comparison were limited to manufacturing because the old nationalizations had been predominantly in fuel and power. International comparisons are not very reliable because of varying definitions, and Table 5.1 should be read as indicating an order of magnitude. It suggests that after this new round, France was in front in the Western European nationalization race, although not very much ahead. But by 1982, the question in Paris was no longer whether the instruments of an alternative policy existed, but whether there was a will to use them.

Table 5.1 Share of Public Firms
(%)

	Value Added	Employment	Investments
France before 1982	11	8	30.5
France after 1982	20	11	35*
Austria	14	13	35
Sweden	14	12	30
Italy	12	11.5	30
United Kingdom	10	12	29

*Excluding the Post Office.

Source: Jacques Blanc and Chantal Brulé, *Les Nationalisations françaises en 1982,* Notes et études documentaires, nos. 4721–4722 (Paris: La Documentation Française, 1983), p. 76.

If France was showing the way in nationalization, it was only catching up in social relations at the level of the enterprise. This is why the Auroux laws can be mentioned briefly. The four laws provided more time during working hours at the disposal of union delegates; obligated employers in plants of more than fifty people to hold at least annual negotiations on wages, conditions of work, and the like (with no compulsion to reach an agreement); mandated greater power for the health and safety committees; and granted new rights for free expression in the factory, which were still to be defined. A fifth law, limited to public firms, stipulated that one third of the seats on the boards of directors be reserved for representatives of the staff. Although there was much pompous reference to "citizenship and democracy in the factory" and to "workers taking their fate into their hands," the alarms sounded by the Right were groundless. The measures, some of which were quite useful, were destined to bring France to the level of collective bargaining and collaboration prevailing in Europe at the time. By no stretch of the imagination could they be seen as foreshadowing a form of workers' control.

Judging by this impressive list of new initiatives, should one conclude that Mitterrand kept all his pledges? Clearly, not all of them, and the nuclear issue is a telling example. Proposal 38 promised a national debate on the need for nuclear power plants, ending with a referendum. In practice, the only concession was a three-month pause, after which the building of nuclear power stations resumed its course. And yet, applying the usual standards, the Socialists did stick rather well to the letter of their program. It is its spirit, the promise to begin a break with capitalism, to initiate a long march toward a different society, that seems to melt into thin air as soon as the Left climbs into office.

As the "spirit" of a project is never easy to define and in the Socialist case it was ambiguous, the handling of nationalization may help us to grasp the point.

In the 1970s, the French Left conceived two functions for nationalization. One was simple and pragmatic. The state had to step in because private capital was not up to it; preoccupied with quick financial gain, it did not invest and modernize enough to be able to stand up to foreign competition. The other was more complex. The Socialists, at least most of them, did not quite say that the nationalized factories would have to produce differently and invent other patterns of consumption. They did maintain, however, that the firms would have to be run not quite in the same way. Self-management, the once-fashionable *autogestion,* was the counterpart to nationalization; without it, the spread of state power involved the risk of totalitarian temptation.

This second interpretation was apparently allergic to office. "From the very first months of left-wing government, *autogestion* disappeared: the word was erased from the Socialist vocabulary."[16] The Socialists were thus left with the pragmatic version, and the emphasis shifted to one of its aspects. Under Giscard and Barre, the argument ran, French big business, complying with the demands of the international market, had specialized and thus had allowed whole sections of the domestic economy to wither away. A major task of the public sector was to fill this gap and to help "reconquer the domestic market." After a time, as planning was clearly being subordinated to the market on the home front and French industry urged to fit the international division of labor, the Socialists no longer had any intellectual justification for nationalization as a means to an end. Significantly, toward the end of their reign, they were accusing the Right of refusing to leave well enough alone and of dismantling the public sector for ideological reasons.

The relative mildness of the factory legislation drives the same point home. Employers were genuinely worried by two points in the original Socialist project. The first was the proposal to set up shop councils, potentially leading to a shop-steward movement and the revival of grass-roots democracy in the factory. The second suggested that the elected factory committee (*comité d'entreprise*) have veto powers over such matters as layoffs.[17] Both were dismissed on the traditional, and irrelevant, ground that a factory is "a place for working." The real reason was that this threatened the hierarchical structure at the work place, and the business community had to be reassured from the start that a Socialist victory did not mean power flowing from below.

"To Govern Differently" was the title of an article in *Le Monde* written by Pierre Mauroy, Mitterrand's prime minister, and is a natural as well as laudable objective for a Socialist government. The snag is that the Left did nothing of the kind. Indeed, the striking feature of its reign from the start was its continuity, the determination to run things in the same way from the very top—where a Mitterrand oblivious to his diatribes against constitutional dictatorship dons de Gaulle's mantle and settles for the part of an elected monarch—to the bottom—where there is no question whatever of allowing the immediate producers to manage their own affairs. Readers may object here to suggestions of fundamental continuity, those who had heard at the time of great tension in France, of one of those stormy, almost revolutionary periods "when cowards flinch and tyrants tremble." My answer is that most of the press, domestic and foreign, was initially hostile to the new regime or, to be more precise, hostile to its radical potential and therefore was inclined to cry wolf. Let me illustrate this point with two examples, most often quoted to demonstrate that at least in the early days of the Socialist regime, France was almost in the shadow of the gulag or the guillotine.

On October 13, 1981, in a parliamentary debate, the Socialist deputy André Laignel said, "You are legally wrong, because politically you are a minority." Taken out of context and raised to the level of a categorical imperative, the sentence does indeed sound ominous. But this was not Immanuel Kant speaking. Laignel was arguing that since the Left now had a parliamentary majority, the other side could no longer impose its legislation on the subject. It may have been clumsy or tautological language, but it was neither revolutionary nor totalitarian, as it was painted.

More quoted still are the words of Paul Quilès, the former manager of Mitterrand's electoral campaign and a future minister, addressing the Socialist congress at Valence on October 23, 1981. The date is significant for the mood. The Socialists were still dizzy with success, although already aware of obstruction. The action provoking indignation at the time was that of the chairman of Paribas selling off its Swiss subsidiary in order to avoid its nationalization, which had been democratically decided by his country. "It would be naïve," said Quilès to his comrades, "to leave in their jobs people determined to sabotage the policy wished by the French people: rectors, prefects, heads of nationalized industries. Nor should we say, like Robespierre at the Convention: Heads will fall. We should say which ones and rapidly." All that was left in subsequent indictments was Quilès, the latter-day Robespierre, and the Socialists as head choppers, whereas if this metaphorical interpretation were applied to the United States, each new administration would be made up of bloodthirsty sans-culottes.

In theory, the French do not have a spoils system, although careers of top civil servants are clearly not unaffected by the political color of the administration. General de Gaulle's victory in 1958 was followed by a gradual purge. The cushy jobs and juicy contracts for the Gaullists led to the coining of a phrase, "the UDR state" (*l'Etat UDR*), based on the initials of the Gaullist party at the time. The election of Giscard to the presidency in 1974 meant that while the Right remained privileged, the bias was in favor of his followers rather than the Gaullists. In any case, public servants with known Socialist leanings had quite a lot of ground to make up.

Concealed to some extent by the quasi-permanence of right-wing rule, a form of the spoils system actually does exist in France, although its victims, being civil servants, are shifted about rather than kicked out. There are quite a number of top jobs whose holders tend to change with the advent of an administration with a different political color. They include, naturally enough, the staffs of the presidency of the Republic, of the head of government, and of all ministers (the so-called *cabinets*, with their *directeurs, chefs,* and other members); the *préfets,* representatives of the government in France's ninety-six departments who are under the direct command of the minister of the interior; the *recteurs,* regional representatives of the minister of education; the chairmen of the nationalized industries and other public bodies (among whom until the Socialist reform were the bosses of television); a few heads of division in each ministry; and so on. All these civil servants can be replaced if the government so chooses. Contrary to expectations, the Socialists, after their long stay in opposition, did not use the discretionary power of dismissal much more than their predecessors. Even more significantly, they barely affected the origins and structure of the so-called power elite.

The French state, and French big business for that matter, have at their disposal highly selected and very skilled public servants. After graduating from high school, the brightest pupils stay on for a couple of years to prepare for special examinations to the *grandes écoles,* which require a highly competitive test on the way in and a ranking examination on the way out. The best known are the Ecole Polytechnique, which trains civilian and military engineers, and the Ecole Normale Supérieure which, while nominally a training college for teachers, used to be the main road to prestige and position. The *normaliens,* however, have been replaced in the corridors of power by the *énarques*—the graduates of the Ecole Nationale d'Administration (ENA), set up after the Second World War. Applicants must have a degree to qualify for its entrance examination, and depending on their rank in the finals, they pick the administrative corps of their choice. On paper, this system based on competitive examinations is perfectly democratic. In reality, society being what it is, it has a very strong bias in favor of children of the professional upper classes. I bracketed big business with the state as potential employers, since many of these civil servants "put slippers on," *pantoufler* being the French term for moving into the private sector in search of much higher financial reward. The importance of the resulting osmosis between public and private need not be emphasized.

Statistical studies of the changes in this administrative Establishment following the victory of the Left show that while many of the civil servants were new, their origin and background were roughly the same as those of their predecessors.[18] The only slight departures from the established pattern were a small increase of the party activists among members of ministerial *cabinets,* although not as a rule in key positions; the replacement of a few public servants with the highest grade (finance inspectors) by those with the grade below (civil administrators), although this did not last; and, in economic jobs, a slightly higher number of people "taking off their slippers," if one may say so, or returning to the administration after a spell in private business. Allowing for these shades, the picture of the "pink elite," as the Socialist Establishment was called, looks very much like that of its predecessor. It may be argued that this is inevitable, that a new government has to find experts somewhere, and that when revolution actually comes, some privileged children betray their class and side with the newcomers. But this was no revolution. The ferment precipitated by the arrival of new Socialist blood did not last. The caste reasserted its rights, and modern mandarins, like their ancestors, are not prepared by their origins, their training, and their privileged position to struggle for a radical break with existing society. Even as modernizers, they are instinctively for continuity, especially when they are given a lead in this direction from the very top.

Ceremonies are sometimes instructive. The Council of Ministers, the official meeting of the government headed by the president, is held in Paris every Wednesday morning. Mitterrand made it plain to his ministers from the start that this is a formal occasion, that they must not use first names but must address one another by their titles. The order of the day is based on an agenda going back to the monarchy, and the only Socialist innovation was to allow "ministers to stand up and fetch a glass of orange juice."[19] Mitterrand shared with General de Gaulle a fondness not so much for pomp as for ceremony, a respect for established institutions and customs of the state. He not only filled the general's shoes, but also obviously enjoyed and loved the inherited system of rule from above, although naturally he used it in his own way.

Mitterrand, as one of his portraitists pointed out, "prefers to slip aside rather than to make the cutting move."[20] He likes time to help him to decide. He gives the same file to three people simultaneously, often unaware of the competition among them. The existing structure of political power suited him perfectly. Its center lay at the Elysée, the eighteenth-century palace that housed at various stages Madame de Pompadour and the two Napoleons and that, since 1873, has been the official residence of French presidents. The big staff is run by the secretary general, and Mitterrand's first appointment, Pierre Bérégovoy, was an exception to the rule about the origin and training of civil servants: He was a Socialist of long standing and not a product of the elite schools; he had started work in a factory at sixteen and then rose to a managerial job at the Gas Board. His successor in 1982, J. P. Bianco, was an *énarque.* Another person worth noting in the room next to the president's was a small bespectacled man, a champion at passing examinations, having come first in three of the top schools, excelling at once in the arts and the sciences, writing possibly too fast for a would-be Pico della Mirandola—Jacques Attali, the president's special adviser and one-man brain trust. Otherwise, the Elysée was filled with the usual experts[21] on matters civilian and military, political and economic, domestic and foreign, allowing the president to keep an eye on all things from a distance and to intervene when he chose.

The Hôtel Matignon, also an early-eighteenth-century building but the prime minister's office for only the past fifty years, the other seat of power, lies on the left bank of the Seine. A lively chronicle by a journalist who served on

Mauroy's staff shows that the prime minister in France is alive, well, and really governing—that is to say, running the affairs of the state on a daily basis.[22] Only he is also his master's servant. There is no delimited "reserved domain" belonging to the president, even if Mitterrand, like de Gaulle, took a special interest in foreign and military affairs. But everything is potentially his province.

To complete our tour of eighteenth-century architecture, the third center of power theoretically lies in the Palais Bourbon, the seat of the National Assembly, just across the bridge from the place de la Concorde. More accurately, since the Socialists had a comfortable majority in that chamber, what mattered to Mitterrand was the undisputed control of his party, which he had left in the hands of one lieutenant, Lionel Jospin, still youthful despite his graying hair, and of its parliamentary group, headed by another faithful, Pierre Joxe, son of a Gaullist minister but himself on the left wing of the Socialist party.

There was no doubt about the apex of this triangle. Jospin, the party secretary, was to quarrel with Laurent Fabius, the next prime minister, over the leadership of the electoral campaign. Tension between Joxe and Mauroy was permanent. Once Mitterrand took sides, however, the verdict was final. Indeed, all the people just mentioned were invited for breakfast at the Elysée every Tuesday with the president and his two main assistants. They were also the most frequent guests—together with Louis Mermaz, the president of the National Assembly; a few ministers; and a few other party leaders—at the Wednesday lunch following the Council of Ministers. They were consulted on tactics, occasionally on strategy, but Mitterrand's ascendancy was comparable with that of General de Gaulle.

How this pyramidal power descended from above is seen

even better by looking at the National Assembly. The rise of the Left was traditionally expected in France to usher in the "professors' republic."[23] The 1981 victory, we just saw, consolidated in the executive the reign of the technocrats. This was not true of the legislature. Over half the Socialist deputies in the National Assembly came from the teaching profession. Although not quite the bearded weirdies painted by a hostile press, most of them shared a lyrical illusion and the belief that they had been elected to accomplish historic changes in French society. Very quickly they learned to march in step, vote to order, and altogether show a discipline worthy of de Gaulle's old faithfuls. Did they never rebel? They tried. In October 1982, President Mitterrand decided that it was good for the nation, and electorally profitable, to let old wounds heal and twenty years after the end of the Algerian war let bygones be bygones. The proposed total amnesty covered the military ranks and privileges of generals who had tried to topple the Republic and then had approved terrorist acts in Algeria and at home. For Socialist deputies in their early forties who, like Joxe himself, had started political life struggling against the war in Algeria and the very people now being rehabilitated, this was like spitting on their own past.[24] They asked for the generals to be excluded from the amnesty. Mitterrand would not hear of it. The government made a purely formal concession: Mauroy used the convenient constitutional device (Article 49, Paragraph 3) by which a bill becomes law if opponents do not carry a motion against it. On the night of November 23, 1982, Socialist deputies getting home from the chamber could tell their families and friends that they had not voted for the shameful law. Whether they could look at themselves in the mirror was another matter and depended on their capacity for self-deceit.

So, power flowing from above everywhere, the Socialists did not govern in a way fundamentally different from that of their predecessors. Did it matter so much? After all, the picture painted in Chapter 3—the ideological subordination and the absence of a dynamic movement and of a conception to deal with the economic crisis—conveyed the vision of a Left doomed to defeat as in a Greek drama. There is a difference, however, between the cool calculation of odds and the resigned acceptance of fate. Life is always more complex. Victory and the resulting euphoria did open up new vistas. On the morrow, a president, a government, a party could have addressed their own electorate, telling it the sober truth: We are keeping our word, granting the increases in wages and social benefits we have promised, but the situation is even worse than we thought—our industry is not competitive; the world economy is not recovering; the balance of trade is running into the red. Let us get together and see what we can do about it all.

A Socialist government may then have been in a position to ask people for sacrifices. The workers were probably ready to tighten their belts for a time, if told for how long and for what purpose and if shown the road ahead. A vision and a project were also the only means for splitting the middle classes. All this, however, required not the pretense of an artificial consensus that concealed contradictions, but the admission of conflict and the search for a homogeneous alliance capable of long-term action. It also required another form of government: not gifts from above and decisions reached in Elysean secrecy, but open discussion and debate at all levels—in factories and offices, in local and ministerial councils—over such key issues as the drawbacks of protectionism and open frontiers, the necessary reshaping of the economy and its social cost, the degree of egalitarianism acceptable at this stage, the frontiers between labor and leisure, democracy in the work place and beyond.

On being elected, Mitterrand presented himself, naturally enough, as the president of all the French. This was rapidly taken to mean a matchmaker, a man of reconciliation and compromise, whereas it could have been interpreted as the champion of one side, presenting the interests of the labor movement as the superior interests of society as a whole. Although a social movement cannot be invented artificially, efforts can be made to encourage its revival. The Right needs only the votes of its electorate. The Left, if it wishes to carry its reforms, requires the active support of its electorate to force the capitalist Establishment to yield. Mitterrand is too astute a politician to have ignored this elementary truth about the necessary balance of forces. If he did nothing about it, he did it consciously.

The failure of the Left in office to weave its many measures into a coherent whole and to present it as an alternative policy was not accidental. The sudden amnesia over *autogestion,* the downgrading of planning, the appointment of the reassuring Delors as the man in charge of the economy, and the decision not to grant the workers any veto power in the factory are all part of the same pattern. The refusal of the Socialists to govern differently from their predecessors was not due to Mitterrand's predilection for royal rule. All these were signs—signals to Washington and Wall Street, to France and the world at large—that the new regime was merely another government, that even with Communists in office it had no intention of crossing clearly defined limits—in short, that it was not a serious threat to the established system.

Should one conclude that from the start the Left in office, while keeping individual pledges, was betraying its promise?

This first year may better be defined as one of wasted opportunities, and the word *misunderstanding* is probably more accurate than *betrayal*. The program of the French Left, after all, was already ambiguous in opposition. The Socialists talked at one and the same time as radicals and mild reformers; they proposed to "break with capitalism" and to make it run better, meaning only the latter in earnest. The snag is that by the time they reached office, the era of reforms without tears was over. A more radical posture was needed, and they were unprepared to assume it. Faced with the rugged reality of a world economic crisis without the active support of a mass movement, they were disarmed. The elected monarch was naked and had to retire and return in a new disguise.

Euphoria, as Mitterrand had warned, was short-lived. The "state of grace" did not last even a year, as the Socialists realized by watching opinion polls and ballot boxes. In January 1982, four special elections to fill parliamentary vacancies were held in constituencies where the electoral results of the previous June had been quashed because of alleged irregularities. In all of them, the Right recovered the seats it had lost. Even if these were marginal and won at the height of the tide, here was a sign of a turn or, at best, a warning. It was confirmed two months later when the French were again called to the polls to elect councilors in half of the country's districts, or cantons. This time, the Left and the Right came neck to neck, whereas six years earlier in the same areas the combined Left had clearly been ahead. There were also indications that conservatives were now coming out to vote and leftists staying at home. They were soon to have additional reasons to do so. In June 1982, the second devaluation of the franc—the first, in October 1981, had been unobtrusive and ineffective—was coupled with a freeze of incomes and prices. A policy of austerity was there in the making, although still perceived as provisional. The eyes were now fixed on the Ides of March 1983, and indeed it proved to be a momentous month. First, the Left suffered a serious defeat by losing many towns in local, municipal elections. Then, a third devaluation was followed, after a dramatic confrontation, by a definite switch in economic policy. Before we examine the resulting ideological somersault, we must have a look at this economic retreat and its political consequences.

6

The Retreat

> Eternal summer gilds them yet,
> But all, except their sun, is set.
> Byron, *Don Juan*

Fortune is supposed to favor the bold, and, imagination not having seized power in Paris, French Socialists may not have deserved any luck. Whatever the truth of such proverbial wisdom, they had none with their economic forecasts. On paper, the scenario they had written on the eve of taking office looked virtuous and realistic. According to international experts, the Western economy was to quicken pace toward the end of 1981, driven forward by its American engine. This was a good world context for a dose of Keynesian deficit financing at home. The initial boom in consumer goods could thus spread to both investment goods and export industries. The Socialist advisers allowed for a slight increase in French output, with unemployment still rising, in 1981 and for a big jump in production, with the number of jobless going down, in 1982. Higher output and tax revenues were thus to help narrow the budget deficit, while rising exports were to close the trade gap.

It looked perfect on paper; only things just did not work that way. The international pundits were wrong. The depression was not yet over, and foreign demand for French goods, instead of growing, declined, while competition in foreign markets strengthened. The American boom did not materialize on time, but the value of the dollar rose by more than half, from an annual average of 4.22 francs in 1980 to one of 6.57 francs in 1982. This did not help French exports enormously, the American market accounting for less than 8 percent of French shipments abroad. It did raise considerably the country's import bill, since a good proportion of French imports, particularly oil, was calculated in American currency. On the domestic front, too, reality did not come up to expectations. Paradoxically, the Socialists had overestimated their heritage, the efficiency of French industry. They had assumed that it had more capacity and competitive edge than it actually did. In an unsheltered economy, this meant that a boost of French consumer demand partly helped to spur the German recovery. Altogether, the growth of French output was much smaller than had been hoped, unemployment did not fall, the budget deficit failed to close, and the gap in foreign payments widened. The Keynesian policy began to be reversed after less than a year and was completely abandoned after twenty-two months.

The Keynesian drive did not last, nor was it ever overwhelming. Let us repeat that the fiscal stimulus, the expansion through budgetary means, amounted to only 1.7 percent of the gross national product over the first two years and could have been absorbed easily under other circumstances. Besides, this Keynesian reflation did bring *some* results. To begin with, France fared better, or rather less badly, than its European neighbors. In the first two years of Socialist rule, the French gross national product rose by 2.6 percent, whereas the German dropped and the British stagnated. By 1983, unemployment in France was a third higher than in

1980. In Britain over the same period, it doubled, and in Germany it rose even faster (at 8.2 percent of the labor force, the unemployed in Germany actually caught up with those in France, although not with those in Britain, where the proportion was 11.1 percent). The trouble with the experiment was that the yield did not correspond to the investment, and the international cost rapidly became prohibitive. François Mitterrand had to change course quickly, whereas Ronald Reagan could have gone on much longer because France did not have the same room for maneuvering as the United States. It is much more dependent on foreign trade, does not dominate the world economy, and does not have a reserve currency. If comparisons are to be made, they should be drawn with Great Britain.

The other point to keep in mind is that the bad luck, undeniable, was also a marginal factor. Even if its expansion had not gone so much against the prevailing trend, the Socialist government would have had to face its crisis. Any left-wing government suspected of temptation to break the established rules must face it sooner rather than later. The program it works out in opposition is outdated by the time it comes into office, because the suspicious businessmen have in the meantime altered the economic score. Pierre Mauroy was astonished when, in his first months as prime minister, he talked to French employers about, say, tax deductions or credit incentives, and all they wanted to know was "What do you think of profit, the break with capitalism, the authority of the boss in the enterprise?"[1] His surprise is surprising. As long as they are not reassured, capitalists big and small take their precautions. They pause for a while before investing and move their money to a safer place. Until the newcomer has given sufficient guarantees.

France was no exception. Indeed, the French Left tended

to frighten more than its policy warranted. At home, as we saw, it was often the case of a velvet hand in an iron-looking glove. Abroad, however respectfully the new government behaved, it was not perceived as really respectable. Its anticapitalist pronouncements were still too recent for that, and the old ghost was being revived by some new moves, such as the nationalizations. The unfortunate timing merely accelerated the inevitable trends. The big trade gap, the rate of inflation higher than in Germany, and the fairly logical deduction that the franc at some stage would be devalued in relation to the mark justified movements of capital, and the operators felt better when seen doing it for financial rather than political reasons, as though the two could be disentangled. Finally, for a Socialist government to declare an all-out war on speculation would have involved entering into a conflict with the international financial Establishment, a confrontation it was making frantic efforts to avoid.

Dismissed contemptuously in May as wrong and unpatriotic, devaluation was thrust on the Mauroy government less than five months later. In the second half of September, it tried to defend the franc by tightening exchange controls and raising the bank rate. By October 4, as part of a realignment within the European Monetary System, the German mark, accompanied by the Dutch gilder, moved up by 5.5 percent, while the French franc, followed by the Italian lira, went down by 3 percent. The whole operation was both unobtrusive and ineffective. In France, with the government just getting into stride, it passed almost unnoticed. Abroad, speculators treated it as a stopgap, the 8.5 percent devaluation of the franc vis-à-vis the mark barely exceeding the annual difference in the rate of inflation between the two countries.

The second devaluation, heralded by a surrealist show,

was quite a different matter. It was the turn of Paris to act as host to the industrial Big Seven of the Western world. Miterrand borrowed the Sun King's palace for the occasion. On June 4, 1982, Reagan, Margaret Thatcher, Helmut Schmidt, and company arrived in Versailles for a weekend of work and festivities. A dinner for 200 in the Hall of Mirrors, a saxophone concert at the Fountain of Neptune, a performance at the Opera, a funfare of Horse Guards along the Grand Canal, a display of fireworks—the Socialist president of the French Republic was giving his guests a really royal welcome, with the French people peeping in thanks to television cameras. Alas, the majestic display did not induce Reagan to go beyond the platitudes about the desirability of reducing currency fluctuations to a reasonable level, nor did the glories of yesteryear restore the current credit of the franc. While the French president was entertaining in the unreal atmosphere of Versailles, his prime minister and colleagues were busy preparing an austerity package that was to give some sense to the impending new fall of the French currency.

Indeed, on the following Saturday, June 12, another important reshuffle of European currencies was duly announced. While the mark and the gilder moved up by 4.75 percent this time, the French franc went down by 5.75 percent. The Germans accepted this big shift only on condition that the French promise to mend their ways. "The agreement that we reached concerning the exchange rate of the French franc," stressed the joint statement, "took into account the important program that the French government intends to implement." On Sunday in Paris, the cabinet put the last touches on the anouncement. The French public was in for a shock. All prices and incomes, with tiny exceptions, were to be frozen by decree for the next four months. The authorities simultaneously promised to keep the fiscal deficit below 3 percent of the gross national product—that is to say, in practical terms no longer to expand but to cut the budget deficit—and, last but not least, to balance the even more important accounts of social security. The fiesta was over.

The crucial turn in the story of the regime was not signaled to the public in the Churchillian tones of "blood, toil, tears, and sweat." The prime minister's message was toned down on orders because the president had not made up his mind whether the switch in policy was to be permanent or provisional and, possibly, did not quite grasp its full implications. The reshuffle of the government at the end of that month, on June 29, illustrated both the extent of the shift and the remaining hesitations. The break with the past was marked by the departure of Nicole Questiaux, the senior woman in the cabinet and the head of a department of social affairs symbolically renamed by the Socialists the Ministry of National Solidarity (it covered social security, the family, and the aged). Critics were holding against her the statement that she "was not a minister of accounts." This obviously did not mean that she was unable to calculate costs and expenditures; she had just been forced to raise the national security contributions. Questiaux was presumably trying to convey the idea that the new regime involved a new logic, in which social need took precedence over abstract cost accounting, and social justice and the long-term benefit over immediate profitability. To replace her, Mitterrand dispensed with the services of his own secretary general at the Elysée. Slightly unctuous, Pierre Bérégovoy was an able negotiator, likely to get on well with labor and employers' unions. Whatever his previous views on the subject, he was put in charge of the vast Ministry of Social Affairs and National Solidarity, as it was again renamed, to balance the books of the welfare state. However efficient the cost cutters, the operation had to affect the quality of the services.

If Mitterrand had opted for austerity, he was not yet clear whether the policy must be applied in a financially orthodox version preached by a Jacques Delors, and this indecision is reflected in the other important permutation. The departure of Pierre Dreyfus from the Ministry of Industry was not a political act; the former head of Renault had joined Mitterrand's government at the age of seventy-three to help him during the transition. The takeover of Industry by Jean-Pierre Chevènement, already a minister of research, was such an act. Mitterrand had a soft spot for the conspicuously ambitious politician who had helped him to win intraparty battles at Epinay in 1971 and at Metz eight years later. A man fascinated by productivity and expansion, Chevènement had contempt for both the advocates of zero growth and the opponents of nuclear energy. Passionately nationalistic, he often sounded like a Gaullist in disguise. The merging of research and industry—into a sort of French embryo of Japan's Miti—in the hands of such a renowned interventionist could be taken as an attempt to combine austerity with the reshaping of the economy under state supervision. Mitterrand seemed to be trying to keep his options open.[2]

He badly needed an alternative, since his personal stock was crumbling. The economic U-turn had immediate political consequences. True, the euphoria was already over, as was shown by the electoral results in January 1982. But the fall from grace was predictable, and the drop in support did not have to last. The loss of popularity that summer, reflected in the sharp decline in the approval rating of the president as well as his prime minister in opinion polls, was more serious. The Left's own electorate saw its hopes dashed and was given no explanation. The Right took advantage of the changing mood to launch an offensive. Not that it had been idle up to then. The French Right, so long in office, had come to consider the state as its property. Stunned by the successive blows of presidential and parliamentary defeat, it quickly recovered enough to start a propaganda battle against the "usurpers" with the help of the press. In France, like elsewhere, the press is predominantly conservative. The Right could rely in particular on *Le Figaro* and the other papers belonging to Robert Hersant, whose youthful collaboration with the Nazis did not prevent him from becoming France's biggest press lord.[3] His papers were now making regularly apocalyptic forecasts of galloping inflation at home, international bankruptcy, and bureaucratic collectivism just around the corner. Such a catastrophic vision made it easier for the Right not to appear as the defender of the privileged few.

Even when urging the many to tighten their belts, the Socialists in this early period were insisting that the wealthy must make their contribution. They rendered the income tax more progressive by raising the top rate to 65 percent. They slightly increased death duties on large estates. They also introduced into the French fiscal system a wealth tax, albeit a modest one. In the process of preparation, productive equipment and works of art were removed from its incidence, reducing it to a sort of real-estate tax on people who declared property worth 3 million francs (about $455,000 at the rate prevailing in 1982, the first year in which the tax was levied). Altogether, the Socialist government did not begin a general reform of a tax system that was regressive even by capitalist standards because it relied mainly on unequal indirect taxation rather than on the more progressive income tax.[4] The taxation measures, judging by official statistics, hurt only a very small proportion of the population.[5] Under the circumstances, it was politically wiser for the opposition to wage a general, catastrophic assault on the government's economic policy. These farfetched accusations in the conservative press and opposition speeches had the additional advantage of putting the Socialists on the defensive, forcing them to deny more radical intentions and, ultimately, driving them into the middle of the road.

The modification of mood in 1982 led also to a variation in the occasional occupation of the Paris pavement by demonstrators. Labor unionists, although disappointed, were not yet ready to march or protest against the government they had helped to elect. The streets were thus ready for the opposition, and the French papers gave a good example of the unsubtle transmutation of values. For the respectable press, strikers and demonstrators had been the scum: laborers who were sabotaging the basic "freedom to work," privileged public employees who were ungrateful for the security of tenure, and pampered students who should know better. Suddenly, the demonstrators became the salt of the earth: heroic fighters who were defending the freedom of a civil society threatened by the encroachments of a potentially totalitarian state. The ideological alchemy, however, is not enough. People are still needed to do the demonstrating, and the Right could traditionally find them mainly among peasants.

Politically dominant in the countryside, the National Federation of Farmers Unions (FNSEA) had developed, under the Fifth Republic, close ties with the ruling Right, which got into the habit of appointing former leaders of the federation to ministerial jobs.[6] Its participation in a complex network of controls over credits and subsidies gave the FNSEA a powerful position in the countryside. Edith Cresson, the energetic Socialist minister of agriculture, tried to break its strangle hold. In the elections to the agricultural chambers (bodies supervising development in the countryside), held in January 1983 for the first time under proportional representation, other unions questioning the line of the FNSEA did capture around one third of the votes cast. It might have been a beginning. It had no sequel. Like so many other efforts to break with routine, this one was soon given up. Michel Rocard, who replaced Cresson as minister of agriculture, tried to collaborate with the biggest organization that is today the FNSEA, which did not prevent the latter from acting as the spearhead of the opposition. The first protest by farmers against the new regime took place as early as March 1982. Many demonstrations and riots were to follow throughout France.

The middle classes, as usual, felt squeezed by the crisis. Even the doctors, whose numbers had climbed from 40,000 to 120,000 in 20 years, had ceased to feel comfortable. The fact remains that this middle-class discontent found an active and directly political outlet only when the Left got into office. The streets of the capital were thus to witness some rather unusual marchers with white collars and ties, waistcoats and umbrellas. On September 20, 1982, doctors, lawyers, and architects demonstrated with masseurs and driving instructors under the common banner of liberal professions, and they numbered close to 50,000. Two weeks earlier, shopkeepers had marched with employers under the auspices of the National Union of Small and Medium Industry, and two weeks later, shopkeepers joined artisans in another demonstration, each time the attendance to be counted in tens of thousands.

When in the spring of 1983 a few right-wing students took to the streets, wishful thinkers within the opposition began to dream of a May 1968 in reverse. More was to come. On June 3, after two policemen had been shot, their right-wing colleagues, joined by Jean-Marie Le Pen and his supporters, marched to the Ministry of Justice in the elegant place Vendôme, where they yelled against Robert Badinter with what looked like a fascist salute. They actually went on, unperturbed by their fellow police officers on duty, almost the whole way to the presidential palace. When in the following February, truck drivers, angered by a conflict in the Alps close to the Italian frontier, blocked many a French road with their vehicles, the allusions to Chile and to the late Salvador Allende were inevitable. They were also a figment of the imagination. The movement of truck drivers, unpopular because it was disturbing vacationers, fizzled out. Those directly responsible for the police demonstration and for the ease with which it proceeded toward the Elysée were sanctioned. Right-wing students were not even a shadow of the May movement. The middle classes did not have the social weight of the industrial workers, their capacity to bring the country's economy to a standstill. Above all, there was no need for subversion because the Socialists made it quite plain that they had no serious intention of shaking the established order. The French events simply suggested how a strategy of tension might have been put into practice had the Establishment felt that its vital interests were in danger. They also taught the French Right how to stage mass demonstrations and use them as a political weapon, a lesson it was soon to apply successfully over the issue of Catholic schools.

Meanwhile, in 1982, the opposition discovered a way to win votes, a terrain on which the Left was vulnerable. To conquer this ground, it was necessary to pander to the fears of an insecure population and to the prejudices of those who are suspicious of anything different, strange, alien. Like their counterparts in other countries, the French conservatives began to parade as the champions of law and order. But here they could also claim to be the defenders of natives against the foreign invasion.

The number of town dwellers rose exceptionally fast throughout the postwar years. Housing, education, the social services, collective life, and even the mentalities did not keep pace with this urban growth, and the lag inevitably helped to build up tensions. The economic crisis accelerated the process, precipitating particularly the spread of petty crime—theft, small burglaries, and mugging. French towns had not been turned into Chicagoes of the Prohibition era, and Paris remained incomparably safer than, say, New York. Nevertheless, insecurity was gaining ground and so was popular discontent, especially in the poorer suburbs. As long as the Right stayed in office, all this, however painful, was an international trend and a product of Western civilization. When the Left took over, the reason suddenly became political, Marxist collectivism and liberal laxity taking the blame for the disarray. To show that the curves of delinquency were following their upward course unaffected or that the prisons, far from having been emptied by the permissive Left, were terribly overcrowded was of no avail, since the indictment was fundamentally irrational. Badinter, having abolished the death penalty and expressed his preoccupation with the prevention of crime, as well as the reintegration of former criminals into society, was, it goes without saying, a lover of criminals full of contempt for the victims. The Socialists, however, were losing both ways because while the minister of justice was pleading for a liberal, reformist interpretation of the rule of law, his colleagues who dealt with the police were preaching law and order in a manner reminiscent of that of their predecessors. More generally, the success of the Left in the field of justice, as a rule, stands or falls with its capacity to alter society at large.

It is easier to find scapegoats than solutions, and the immigrants were an ideal target. For the opposition, it was just a case of confirming the virtues of selective amnesia in politics. The Right simply had to forget that the mass migration of the 1960s and early 1970s had been carried out during its political reign and had been inspired by French employers seeking cheap and flexible labor. Once the extent of the economic crisis was perceived, strict controls were introduced and the flow of immigrants was reduced to a trickle. Between 1975 and 1982, the share of foreigners in the French population rose by a fraction, from 6.5 to 6.8 percent. The Socialists, having allowed illegal immigrants residing in France to regularize their position, did not open frontiers or reverse the trend. The crying over the vicissitudes of a large immigrant population was done by politicians responsible for its existence.

Foreigners were convenient targets provided that one told only half the story, or, to use a newly fashionable euphemism, one were "economical with the truth." Did not France have 2 million immigrant workers and just as many unemployed? Were not foreigners a burden for social security, taking up seats in school and beds in hospitals? Were immigrants not more than proportionate among petty criminals? All these rhetorical questions lose their impact if they are answered otherwise than by jingoist monosyllables. Thus the sharp rise in unemployment in France after 1975 coincided with the end of mass migration. Besides, foreign workers do not as a rule fill the same jobs as native workers, and the idea of simple substitution is an absurdity. The immigrants arriving mostly as adults, France actually saved on their upbringing, and, on balance, they were not a drain on social security. Finally, the rate of criminality is not higher among foreigners than among French natives of the same age and social group (although *petty* crime is naturally more prevalent among the jobless youth of the suburbs than among the industrialists and bankers living in the posh districts of Paris). All these facts and many more were to be found in a useful booklet that the government printed in 2.2 million copies and then scrapped at the last moment in March 1983. On the eve of an election, the less said about the matter, the better—so concluded the Socialists, lacking the courage of their convictions.

The Left was paying the price for past sins and current lack of courage. It had not struggled consistently against racism and xenophobia amid its supporters while the going was good. By the time the crisis and unemployment offered scope for antiforeign feeling in general and Arab-bashing in particular, it was on the defensive. The judgment applies to the Left as a whole. The Communists, proclaiming on their banners that workers of all lands should unite, forgot this principle in the Red suburbs, their own electoral fiefs. They gave the impression for a while of hoping to avoid the backlash by taking a tough line on the immigration issue.[7] Socialists were hardly in a position to play holier than thou.[8] True, the left-wing government, to begin with, did a few things to improve the lot of immigrants: allowing illegal ones to settle, lengthening the residence permit from three to ten years, extending the right of association. But the pulping of the booklet was not an isolated instance. It was over immigration that the government showed its loss of nerve very early. In August 1981, on a visit to Algiers, Claude Cheysson, the French foreign secretary, suggested that France keep number 80 of Mitterrand's 110 proposals and grant foreign workers who had lived in France for some time the right to vote in local elections. The French Right thereupon raised a pandemonium, and the government dropped the proposal, which was never taken up on the ground that the country "was not ripe." As though the task of the Left was to lag, not to lead.

The center of gravity had shifted rightward. The Left, attempting to reconcile its conscience with its electoral preoccupations, was thrown on the defensive. The respectable Right, which used to ignore the issue as long as some of its financial backers required foreign labor badly to make their profits, chose to wield it now as a club to bludgeon the "usurpers." However unscrupulous in their choice of weapons, the orthodox conservatives were to find competitors ready and able to stoop lower than themselves. In rendering racism almost honorable, they had worked for Le Pen and his National Front. They prepared the ground in France for a revival of the extreme Right.

"If you want a nigger as neighbour, vote Labour" was the theme of a Tory whispering campaign in a notorious British special election. A more polite version of the refrain, putting the blame for all evils on foreigners in general and on Arabs in particular, was to be heard in most French towns as the Right began building up its own electoral campaign. The municipal elections of March 1983, since they involved neither president nor parliament, were not supposed to affect the country's policy. The opposition was determined to use them not only to reconquer control over a number of towns, but also to keep the Left on the run, to prevent it from carrying out its program on the ground that it no longer had a mandate to govern, let alone to accomplish structural reforms.

In the second half of 1982 and at the beginning of 1983,

two battles were actually being waged at the same time. One was open and preelectoral. The other, fought behind the closed doors of the Elysée Palace, was more momentous, since it concerned the economic policy of the government and, therefore, its global strategy. Despite the relative success of the wage and price freeze—the cost of living in 1982 rose by just under 10 percent—and the $4 billion borrowed from foreign banks, the franc was under permanent threat and other measures to restore the economy were clearly required. Mitterrand was busy consulting. Austerity was not the issue, only the context in which this policy was to be applied. During office hours, he heard the arguments of the government, the Delors line of financial orthodoxy within the European Community, which was increasingly resembling the policy of Raymond Barre, Mauroy's conservative predecessor. The *visiteurs du soir* ("after-hours guests"), as the prime minister called his opponents, were preaching an exit from the European Monetary System, a spell of relative protectionism, more intervention by the government in the economy, and less subservience to international finance. The opposition between the government and the evening visitors is not quite accurate because among them were such ministers as Pierre Bérégovoy, Laurent Fabius—the favorite son groomed for higher things but presently in charge of the budget—and Jean-Pierre Chevènement. There also were outsiders, such as Mitterrand's personal friend Jean Riboud. This former resister, deported to a German concentration camp, after the war rose in the business world to become chairman of a French multinational, Schlumberger. The two battles, the electoral and the economic, were intimately linked, although for narrative purposes it is better to deal with them in succession.

Municipal elections are held in France every six years.

The town councils, and thereby the mayors, are thus chosen simultaneously in over 36,000 communes, from small villages to the capital. These local elections had gained in importance because of decentralization, but they had a clear political significance mainly in the 221 larger towns of over 30,000 inhabitants. On the previous occasion, in 1977, the Left had captured 60 more such towns, bringing its total to 155. This, however, had been a period when the Left was at its most dynamic, with the Communists still faring fairly well and the pink Socialist wave spreading, invading in particular the formerly conservative regions of Catholic western France. Some loss was inevitable, although the consequences depended on its size. Thirty towns, half the gains of the Left, was considered by the government as a defeat from which it could hope to recover; more than that would spell disaster.

Taken aback by the policy of austerity, the left-wing electorate was then bewildered by the total absence of an explanation and a perspective. Even the immediate line of the government was uncertain. On January 31, Edmond Maire, the leader of the progovernmental French Democratic Labor Confederation (CFDT), emerged from the Elysée Palace and, having just seen the president, warned that "a second austerity plan may become necessary." On February 16, appearing on television, Prime Minister Mauroy told the French people that "all lights were turning green."[9] While the Left thus did its best to discourage its followers from voting, the Right mobilized its supporters through an effective campaign centered around insecurity and immigration. The result exceeded governmental fears. In the first ballot, on March 6, the swing was quite impressive. Limiting the comparison to big towns, the Left, which had mustered 53 percent of the vote in 1977, got only 46.7 percent six years later; the share of the Right climbed from 44.8 to 51.7 percent. Although an absolute majority was required to win outright in this ballot, the Right captured sixteen large towns. The Socialists lost eight, including Grenoble, whose conquest in the 1960s symbolized the success of the Left in winning over the new middle classes; Brest and Nantes, part of the recent westward drive; and Roubaix in the industrial north, where a Socialist first became mayor seventy years earlier. The Communists lost eight towns, too, among them Arles in the south and several suburbs in the Red belt of Paris, which was beginning not to deserve its name. And much worse was to come. If the position of Mauroy in Lille looked only difficult, that of Gaston Defferre in Marseilles or of Chevènement in Belfort seemed hopeless.

Then, between the two ballots, the Left struck back. It no longer pulled its punches. It called a spade a spade and a racist campaign—stinking. Carried away, a minister even referred to the "fetid breath" of Jacques Chirac. At this point, the press, which hitherto had not been shocked, began screaming about intolerance and the unbearable climate of debate. Presumably, the jingoist campaign, putting the blame for everything on wogs and wops who ought to be shipped back across the Mediterranean, had the smell of Chanel No. 5. Found aggressive by all those who were pushing the government toward consensus politics, the campaign had its effect. It awakened part of the left-wing electorate and sent it to the polling stations. Naturally, this was not enough to reverse the trend. The Left was dispossessed of another fifteen large towns. The Communists lost their southern fiefs of Béziers and Sète as well as their recent industrial conquest, St. Etienne. The Socialists were defeated at Carcassonne, Cherbourg, and St. Malo. But the prime minister won comfortably, while Defferre and Chevènement were saved by the skin of their teeth. The defeat was undeniable; the disaster was finally avoided.

This has its immediate consequence on political choice. After the first ballot, Mitterrand had decided to change his line—opting for the *visiteurs du soir*—and his prime minister. After the second, he chose to keep the man and alter the policy. On Monday March 14 in the morning, before going to see the president, the prime minister read a long piece in *Libération* foreshadowing a change of government and outlining the "dynamic alternative" based on France's departure from the European Monetary System. The article must have been inspired by leaks from the Elysée, since this was roughly the program that Mitterrand presented to Mauroy. With one major difference. He asked the prime minister to soldier on. This was when the real surprise occurred. For the first time in their post-1981 duet, Mauroy dared to say *no*. He was not the man for the new policy.

Mitterrand had to improvise rapidly, since the franc was under heavy attack. The first ballot of the French election coincided with Germany's parliamentary poll, in which the Christian Democrats of Helmut Kohl gained an unexpectedly comfortable victory, and hot money was now flowing quickly to Germany. When the president and the prime minister saw each other again in the evening, Mauroy still said that he "can't drive on icy roads," but proposed that the French should bargain with the Germans over the respective rates of devaluation and revaluation first and envisage an exit from the EMS only if the Germans proved intransigent. By Wednesday, the prime minister, having remembered the break with Mitterrand at Metz and reluctant to reopen old wounds, yielded altogether and told the president that he was ready to carry on, whatever the line. By then, however, for Mitterrand the problem was no longer the man, but the policy and his ability to change it. The evening visitors had seemed to be on the ascendant. Their position was now crumbling. That very Wednesday, Fabius consulted the head of the Treasury, Michel Camdessus.[10] His report was gloomy. French foreign-currency reserves, not counting the stock of gold, were down to about 30 billion francs, or just over $4 billion. Allowed to float, the franc was expected to sink, to drop by over 20 percent. Fabius got cold feet and, backed by Jacques Attali, convinced Mitterrand to yield and stay in the EMS.

Paradoxically, this was when the controversy, supposed to be unsolved, hit the headlines of the international press and was used as an instrument in the tight negotiations with the Germans. Delors, the ideological winner, behaved as though he were under terrible pressure. Arriving in Brussels on March 19 for the meeting of finance ministers called to settle the financial crisis, he vituperated against the "arrogant and uncomprehending" Germans. The Germans were worried: If the most "European" of French ministers was so violent and bitter, his opponents must be gaining the upper hand. The morning after, Delors told journalists that "important things will happen this afternoon in Paris," where he was going, and they interpreted his statement as the announcement of his promotion to the premiership. On a flying visit to the French capital, Delors saw the president twice, and the meetings were strained. Mitterrand hates to have his arm twisted. Finally, nothing was decided on the domestic front. When on Monday morning, Mitterrand, in turn, flew to Brussels for a European "summit," Mauroy, who drove with him to the airport, did not know whether he was staying in the government or whether he ought to look for another job. That very afternoon in Brussels, the EMS was submitted to a most comprehensive realignment:

All eight currencies were moving, five up and three down. The mark was revalued by 5.5 percent, and the French franc was devalued by 2.5 percent.

On Tuesday, back in Paris, the president invited Bérégovoy, Fabius, and Delors to a late lunch. Delors overplayed his hand. He wanted the premiership and the control of the economy, with neither of his fellow guests at the Ministry of Finance. Mitterrand is not a man to appoint a potential competitor. When Mauroy arrived at the Elysée in the afternoon with his letter of resignation, he was able to keep it in his pocket. Just before midnight, the composition of the third Mauroy government was revealed, and it brought few surprises. Delors, in charge of the economy and the budget, came second on the list, just after the premier. Bérégovoy, minister of social affairs, came third. The only important departure was that of Chevènement.[11] Having declared that "a minister shuts his trap, and if he wants to open it, resigns," he was now in a position to speak up. The Ministry of Industry and Research was taken over by Fabius on his planned journey to the top. The only other significant move was that of Rocard, who was shifted from the ineffective Ministry of Planning to the awkward Ministry of Agriculture.[12]

The first act of the reshuffled government was to present the bill, and it was stiff. France's 15 million or so taxpayers were asked to contribute 1 percent of their taxable income to cancel the deficit of social security. Nearly half that number, those better off, were forced to lend the state for three years 10 percent of the income tax they paid that year. If you add to it higher taxes on gasoline, alcohol, and tobacco, increased prices for gas, electricity, and transport, and budget cuts, the deflationary package came to 3,470 billion francs, or roughly 2 percent of the gross national product. In other words, it was about twice as big as the original inflationary dose prescribed by the Socialists two years earlier amid such fuss. This time, the passion of the press, French and foreign, was aroused by only one item—the drastic limitation of the foreign-currency allowance to under $450 per adult. This was described as a really symptomatic step on the Socialist "road to serfdom": The French were already being tied to their soil. Such rhetoric, however loudly drummed, could not attract a mass audience because of very pedestrian facts. Roughly one French citizen in two goes on vacation, and less than one in six travels abroad. Besides, about $1,500, plus fare, is probably as much as many a working-class family of four can afford for its summer vacation on Spain's Costa Brava. Right-wing preachers were not going to be bothered by such trivia: "the working class, if such a thing still exists, often identifies less with its real conditions than with its aspirations. So when a trip to the United States or Katmandu becomes next to impossible, the government is seen as padlocking the dream factory."[13] Let my people dream—the new slogan opened up infinite possibilities for justifying the privileges of the leisured classes.

But let us get back from this world of fancy to the grim reality of Socialist rule in time of crisis. The ten days or so around the Ides of March were described here in such detail because they finally clinched the economic policy of the Socialist government and, therefore, its political fate.[14] True, the U-turn had begun earlier, and the austerity course had been on since the previous June. Besides, even now, with the line confirmed, the president addressed the nation on March 23 in undramatic fashion, as though continuity had been the characteristic feature of the reign. Yet for all this, the March confrontation was crucial. Let us first see what this battle was not. The idea that people should be actually involved in this struggle was not advocated by either side; it had been ruled out from the very start. Nor was it a conflict between defenders of the capitalist framework and opponents determined to dismantle it. The *visiteurs du soir* did not belong to the second category. The chairman of Schlumberger, one of the most influential among them, could hardly be described as a crusader against capitalism. Indeed, the mission he conceived for his friend the president was to reconcile the French Left with dynamic capitalism. "It was up to the Left," he argued time and again,[15] "to free the spirit of enterprise and the entrepreneurs, to carry a mental revolution in this field. And this the Right couldn't do, only the Left can. I hope Mitterrand will finally realize that this is his principal task."

The battle was nevertheless important because it was a final attempt to keep France marching out of step, unaligned with Germany, a last effort to preserve a degree of independence, some room for maneuver, and an element of hope or illusion. Probably the effort was condemned from the start once the decision had been taken not to rely on popular support. If the purpose was to move smoothly, without a radical break and without mass mobilization, the advocates of an alternative policy certainly made their bid much too late. By the time the *visiteurs du soir* were supposed to present their final solution to Mitterrand, the international financial accounts were favoring the other side. And so March marks the total conversion. The Left, to quote Jacques Delors, could ultimately get on with its task of "ensuring a better performance of capitalism in France."

A conversion has its consequences and a system, its logic. For the French Left, the conversion now meant a shift of emphasis from solidarity and social justice to private initiative and gain, from equality to profitability, from the struggle against unemployment to that against inflation. The new austerity plan implied a cut in domestic consumption and a rise in unemployment, which could be concealed only by statistical make-up. Within a few months, once Mitterrand had properly grasped the full meaning of the decision, the search for an alternative, the very admission that one could exist, would be treasonable behavior for Socialist party members. If the term *autogestion* ("self-management") had ceased to be used by the Socialists as soon as they got into office, now socialism itself made a discreet exit from the vocabulary of Mitterrand's government.

Was there no risk of the Left feeling betrayed by such behavior? To avoid such disappointment, the government chose to combine the economic withdrawal with a political offensive. The Socialists were going to liberalize radio and television, liberate the press from the strangle hold of financial tycoons, and, last but not least, set up a "unique and lay national service of public education," thus keeping an old pledge and showing egalitarian feelings. The mistake was to assume that the economic and the political could be separated. Entertaining utter confusion over the role of the state and allowing the private to be treated in principle as superior to the collective and freedom to be considered as a product of the market, the Socialists were bound to lose on the three issues of their choice. The economic retreat, in turn, paved the way for the ideological surrender.

7

Cultural Counterrevolution

> I am ashes where once I was fire.
> Byron, "To the Countess
> of Blessington"

Dreux is a sleepy town of about 35,000 inhabitants lying some 60 miles from Paris. In 1977, as the pink wave was spreading westward, the town was conquered; Françoise Gaspard, a young Socialist woman, became its dynamic mayor. Six years later, her list defeated the opposition by merely eight votes. Because of the closeness of the result and of alleged irregularities, the vote was quashed and a rerun ordered. In September 1983, the eyes of France were suddenly focused on Dreux, turned overnight into the racist capital of the country. There were two joint reasons for this unexpected fame: the presence in Dreux of a relatively high proportion of foreigners, about one fifth of the population, particularly North Africans;[1] and the activity of Jean-Pierre Stirbois, the second in command of the National Front, who had been nursing this constituency for years. The general mood of the country after the electoral campaign of the previous spring gave him his real opportunity. On September 4, in the first ballot, the united Left, no longer headed by Gaspard, who had moved down the list to dedramatize the conflict, got only 42.7 percent of the vote; the respectable Right, 40.6 percent; and the National Front, 16.7 percent. The odds were that the honorable Right would win the second ballot without shameful alliances. The local followers of Jacques Chirac, Valéry Giscard d'Estaing, and Raymond Barre decided to take no risks, and since the law allowed mergers, they included in their list of thirty-nine candidates four members and eight sympathizers of the National Front. It was as though the Republican party had given its official blessing to John Birchers or Ku Kluxers. This recognition of open racists as members of the family provoked controversy and passion throughout France. It even mobilized some left-wing voters in Dreux, although not enough to redress the balance. The stake in that election, however, was not the provincial town hall. Dreux, by rendering neofascism respectable, marked the reentry of the extreme Right into French politics.

Jean-Marie Le Pen is the symbolic beneficiary of this trend. No longer wearing a black patch over his left eye, lost in a brawl, red-faced, rather fat and smartly dressed, and saying aloud "what everybody really believes" but barely dares to whisper, Le Pen was enjoying his belated success. Although only fifty-five years old at the time of the Dreux by-election, Le Pen had a long and checkered career that sums up the vagaries of the extreme Right in France throughout the postwar period. The orphaned son of a Breton fisherman, he came to Paris to study law and, thanks to his talent for fisticuffs and oratory, rapidly graduated through student faction to the fascist fringe of French politics. Since nobody likes to be squeezed in the name of economic progress, Poujadism—from the name of its leader, Pierre Poujade, the owner of a provincial stationery shop—was a lower-middle-class, essentially shopkeeper, reaction to concentration in the French economy. The great surprise, in 1956, was the election by this movement of more than two score deputies to the National Assembly. Le Pen, at twenty-seven, was one of them and quickly built his reputation for vituperation and racism. ("You crystalize in your person a certain number of patriotic and almost physical repulsions," addressed by him to Pierre Mendès-France, was a notorious example.)

His main ambition, however, was to contribute to the fall of the Fourth Republic, and the main battleground was in Algeria. Parting company with Poujade, he volunteered as a paratroop officer and served in the repressive forces at the height of torture during the battle of Algiers (hence his reputation, contested in court, as a torturer). Although the French settlers and their military allies did manage to bring the Fourth Republic down, they had their victory stolen from them by the illustrious fellow plotter Charles de Gaulle. When the general, in turn, was forced to seek peace in Algeria, the military barons tried to rebel against him. After a first defeat, they set up the so-called Secret Army Organization (OAS), spread terror throughout Algeria, and even exported it to France. Le Pen, while not concealing his sympathies, was too clever to go underground and join them in an obviously losing battle. He caught public attention once again in 1965, three years after the end of the Algerian war, as the campaign manager of Jean-Louis Tixier-Vignancour, a former junior minister in Marshal Philippe Pétain's Vichy government and chief attorney for the OAS. Tixier, champion of colonialism and collaboration with the Nazis, attempting to unite the mourners of Algérie française with older fragments of French reaction, managed to get 5 percent of the votes cast in that year's presidential poll, the last significant score for the extreme Right for a long time to come. For Le Pen, there followed fifteen years in the wilderness, years of petty maneuvering on the neofascist fringe. His personal fortune miraculously improved with the legally contested heritage from a follower,[2] but the political horizon seemed closed. In the presidential poll of 1974, Le Pen, this time a candidate, gathered 0.75 percent of the vote. In 1981, he could not even obtain enough elected sponsors to run. It took the economic crisis *and* the victory of "Marxists" to give him a chance, which he did not miss.

Commentators tried to minimize the success of the National Front by putting it in a historical perspective. France has had a tradition of anti-Semitism going back to the Dreyfus case and beyond. Before the Second World War, fascism paraded openly, and racism was much more virulent. But this was before the Holocaust. Then, for some thirty-five years after the war, the racists had to conceal their feelings in conventional politics. Now, with the blessing of the respectable Right, Le Pen was rendering racism almost respectable. Probably most worrisome in the situation was not the fact that one person in ten was now ready to vote for a party whose leader was an open admirer of General Augusto Pinochet and apartheid, a man who would soon sit in the European Assembly next to his Italian friend, the neofascist Giorgio Almirante. Having prepared the ground for the National Front, the respectable Right was now likely to pander to the prejudices of its electorate. The resistible rise of Le Pen illustrated a radical shift in French politics, a metamorphosis in the substance and form of the political debate.[3]

We may get an idea of this profound change in mood going back to the case of Dreux. The decision to consider the National Front as part of the family had not been approved unanimously. There were some liberal conservatives, admittedly few, who disagreed publicly. Simone Veil, the former chairwoman of the European Assembly and a survivor of a Nazi concentration camp, said that had she been a voter in Dreux, she would have abstained. Interviewed on a popular radio program, Yves Montand, the world-famous singer and actor, proclaimed passionately that he would have followed her example. But his intervention had exactly the opposite impact. Son of Italian antifascist immigrants—Ivo Livi is his real name—Montand could not vote for a list including the National Front. Besides, his background and his following, unlike Veil's, were linked with the Left. Each time she was listened to by one of her side, it was one vote less for racism; each time Montand was, it was one less against it. And indeed, in his indignant pronouncements, he aligned himself with conservative politicians or thinkers, such as the political scientist Raymond Aron (a Jew like Veil and, therefore, proof that origin, on its own, is not a sufficient immunity), all arguing that four National Front councilors in Dreux were less dangerous than "four Red fascists" (read "four Communist ministers") in Paris. The attitude was eloquent, considering Montand's past.

In the 1950s, Montand was not just a successful singer. He was the jolly prole, the bearer of the future, a symbol standing somewhere below Picasso's dove of peace. He and his wife, Simone Signoret—the beautiful Goldilocks, or *Casque d'or,* from the film of that name, who succeeded while aging to become a genuinely great actress—were the progressive couple, a model for millions. They stood on so many platforms, contributed to so many causes that many people assumed, wrongly, that they were Communist party card-carriers. They were merely fellow travelers who came to feel that they had gone one station too far. In 1956, after much soul searching, Montand decided to keep his singing engagement in Moscow despite the Soviet invasion of Hungary.[4] This was the end of the affair, and they then tried to make up for past mistakes, but with the dignity of people remaining true to their convictions. They both played in *The Confession,* a Costa Gavras picture about the "Moscow trials" in Czechoslovakia, but Montand had also starred in *Z,* the same director's film denouncing the Greek colonels. They kept giving their time, talent, and money to various causes—relief for the victims of Franco in Spain, the Chartists in Prague, the mothers of the "disappeared" in Buenos Aires, and political prisoners in Poland. Then Montand, although not Signoret, began to show a bias. Presumably under the influence of his *nouveaux philosophes* friends, who recognized the asset they had in a man of his charisma, craft, and mastery of the media, he reverted to the Stalinist syndrome in reverse, with the line determined by the existence of the main enemy—the "Red peril" and no longer "American imperialism." Indeed, in the radio broadcast after the Dreux election, accused of echoing the arguments of the Right, he cursed, choking with indignation: "They always throw it at you: 'Be careful, you play into the hands of Reagan.' Shit . . . the principal enemy is not there, he is in the gulag."

Montand was worth quoting not only because his words tend to get a naturally larger echo, but also because he put in crude terms what his new allies prefer to peddle in verbose dissertations. Their common attitude raises the question of the uses and abuses of "anti-Communism" in Mitterrand's France, where an unreformed Communist party gave plenty of real causes for anger and many sticks to beat it with, but where a whole section of the Left needed the party as a villain and the Soviet Union as the devil incarnate to justify its own betrayal of any form of socialism whatever. Poland is a good beginning for tackling the issue.

The saga of Solidarity aroused probably greater passion in France than in any other Western country. Partly because of traditional links, Poland in the nineteenth century was perceived by progressive French as the great sufferer and therefore the potential redeemer of nations, largely because the resurrection of a labor movement in Eastern Europe revived old hopes in the collective unconscious of the French Left, which could also project onto Poland its own unfulfilled myth of *autogestion*. The wrath then measured up to the expectations. General Wojciech Jaruzelski's military coup took place on December 13, 1981, a Sunday. The next day, Paris witnessed an impressive, although spontaneous, protest march. It was a left-wing demonstration, although not one of the entire Left, the French Communist party and the CGT—the labor union dominated by the party—choosing in France, unlike in Italy, to side with Jaruzelski's tanks rather than with the Polish workers. But Gaullists who wanted to join the march were chased away by leftist stewards with the slogan "Go back to Chile," a reminder that they were against one military coup but in favor of another. The letter signed by many intellectuals protesting against Claude Cheysson's clumsy statement that there was nothing to be done could still be interpreted as a form of radical indignation. Thereafter, the attitude became more ambivalent. On reflection, ambiguity was there from the start. A section of the French Left, while welcoming Solidarity as a movement directed against Communist rule, was perturbed by this potential model for change in Eastern Europe. It was more at ease with Solidarity dead than alive. At the funeral, its spokesmen could pronounce orations proclaiming with Jeane Kirkpatrick and Ronald Reagan himself that the Soviet bloc is a permanent hell with no possible exit. Come to think of it, this metaphysical belief in the utter impossibility of change in Eastern Europe is totally incompatible with real collaboration with organizations such as Solidarity, which, if their ambition is to be authentic social movements, must assume that, sooner or later, change is possible. Simply stated, much of the activity in the West allegedly in favor of freedom in Eastern Europe has less to do with Poland, Russia, or Hungary than with domestic preoccupations.

The hypocrisy of Western conservatives who love labor unions provided they are in the Soviet bloc and preferably crushed is well known. The French have added an ingredient, the use of Eastern Europe in the internal struggle of the Left. Those Socialists who had opted for Michel Rocard and a break with the Communist party were compelled after Mitterrand's victory to eat humble pie. By 1982 and 1983, they had recovered. Things seemed to be going their way. In foreign policy, Mitterrand was splendid, since he was tough with the "main enemy" and most helpful to the chief ally. Domestically, after some initial daydreaming, he had come to his senses and to the task of "improving capitalism." There remained only the last logical step to take—to get rid of the Communist ministers. The cowardly behavior of the Communist party over, say, Afghanistan or Poland had fundamentally little to do with it, but it was an excellent instrument in the campaign to ostracize the party. The primary purpose was to push the Socialist party into the center of French politics, to forge an alliance with liberal conservatives, thus sealing definitely the submission of labor to capital. In other words, the aim was to revive the third force between the two extremes, a policy of coalition with the moderate Right practiced by the Socialists under the Fourth Republic. Only nobody wanted to recall the past because the name of the main practitioner of collaboration, Guy Mollet, the man of Suez and the Algerian war, still sounded as an insult to Socialist ears.

This is not a parody of the critics of the Communist party, nor should it be taken as a whitewash of that party's conduct.[5] Simply, the political journey of some of the critics who began by attacking the Communist party from the Left and are now bombarding it from the Right is scarcely believable. Khrushchev's speech and the Soviet invasion of Hungary were not the only reasons for divorce from Communism. A whole generation of radicals broke with the Communist party over Algeria, because of the party's refusal to endorse the nationalist demand for independence. Some of those who between 1954 and 1962 had accused the party of not being sufficiently anticolonialist were in 1983 signing a petition accusing Mitterrand of having intervened too little and too late in Chad, of not acting eagerly enough as Reagan's gendarme in Africa—in short, of not being neocolonialist enough. But the biggest break occurred in 1968, when radical youth accused the party of simply deserting the cause of anticapitalist revolution. A few of those critics have since found their niche in the media as panegyrists of capitalism, but they are still against the Communist party. Indeed, "anti-Communism" seems to be their only link with the past, an artificial form of self-identification.

The intellectual mood in Paris in the middle of Mitterrand's reign was a strange one. A foreigner having left the French capital in 1968 and returning there fifteen years later would not have recognized the place. France was the only country in Western Europe where no genuine debate on the nuclear issue, civilian as well as military, was taking place; Paris, the only capital where Reagan's foreign policy, with its "Euromissiles," was approved almost unquestioned. Indeed, "Socialist" France rather than Thatcherite Britain was the country where Reaganism as a doctrine, if one may give it such a highbrow title, was meeting the least resistance among the intelligentsia. The only confusion for visiting Americans was that their home-bred conservatism was known in France as *libéralisme*—the doctrine of laissez-faire, in theory a return through "deregulation" to untrammeled forms of capitalist competition.[6] Even Marxists, aware of the principle that "the ruling ideology is the ideology of the ruling class," could only be impressed by the speed with which French capitalism had restored its ideological hegemony. We looked at some reasons for this recovery in Chapter 3. Here we want to glance at the expression of this shift in the media, limiting ourselves to the press and, since the evolution of, say, Robert Hersant's empire is easy to guess, sticking to three organs—an evening and a morning paper plus a weekly—known for their objectivity or sympathy for the Left.

Pride of place must go to *Le Monde,* for so long the pride of the profession, proving that a nation's main paper need not be the mouthpiece of the financial Establishment. In the early days of *Le Monde,* its writing was painful prose for the American State Department because of the paper's leaning toward nonalignment, its refusal to take NATO as a godsend and German rearmament as inevitable. By 1983, it had become pleasant reading for the White House. Whatever the subject, come rain or come shine, its foreign leader writers seemed to be repeating, like *delenda est Carthago,* that Europe needs its Pershings and its Cruises.

The morning paper is a lively newcomer, *Libération.* Its original team was made up of young people coming from various factions of the extreme Left, predominantly of Maoist persuasion. Since then, it has become a vivid organ for aspiring yuppies with a difference, a sort of daily *Village Voice*. Its politics has followed the evolution of its editor. After 1968, Serge July was urging revolutionaries to prepare for civil war.[7] By 1983, he was sharing the views of one of the paper's financial backers, Jean Riboud; July, too, wanted to convert the Left to the service of private enterprise. Actually, his paper could have published the editorial that was to grace the columns of *Le Nouvel Observateur,* the weekly of the leftish intelligentsia. The editorial proclaimed in earnest and not in sorrow: "Each era has its model: after Sartre comes Tapie."[8] Since Bernard Tapie was a fashionable tycoon who specialized in "restructuring" bankrupt enterprises, a sort of poor man's Lee Iacocca, the message meant that the time of political commitment was over and that of successful climbing had begun, a sort of French, although not very much more sophisticated version of "I'm all right, Jack." This mood does not rule out compassion, and private charity was actually to enjoy its heyday. But the radical transformation of society—*Libération* and *Le Nouvel Observateur* will make it plain in unison—is good for dinosaurs or for gulag lovers.

A man on the spot, in the know, may often give the inside version of a story. I could thus explain, say, the drift of *Le Monde* by the fact that Jacques Fauvet, its editor for thirteen years, retired in 1982 or that the foreign editor appointed earlier had done a stint as correspondent in Moscow, a painful experience that may account for a great deal. Anecdotal interpretations, however, are irrelevant when the shift is general and unmistakable. Deeper explanations are then required to justify the trend.

The chief reason for the ideological shift lies in a double failure of the labor movement: its incapacity in the 1960s to seek solutions beyond the confines of existing society, and its inability in the 1980s to carry out a progressive policy within the framework of a capitalist society in crisis. The decision finally confirmed in 1983 to revert to classical capitalist methods to deal with the crisis had almost immediate results. Prodded by the government, France got busy "restructuring" its steel and shipbuilding, its engineering and car industries. Since this process involved the closing of plants and the laying off of labor, the rise of unemployment, slowed down for a while, resumed at a quickened pace: In 1984, the number of jobless rose by 16 percent, topping 2.5 million, or about 10 percent of the labor force.

All this did not go without resistance.[9] Angry steelworkers from the Lorraine invaded Paris on April 13, 1984. The Left had once promised the survival of the industry in the region and was now signing its death warrant. Eleven days earlier, during his press conference, a Mitterrand just back from the United States and full of high-tech and Silicon Valley jargon, confirmed the verdict, promising compensation for the areas laid to waste. There is no question that the workers were not ready to mobilize fully against their "own government" and were accepting from it measures—for instance, the suppression of the guarantee that wages must keep up with prices—that they would not have accepted from the Right without a bitter fight. But resignation should not be confused with fondness. While the Socialist government was getting its blessing from the Bourse—the shares in Paris, after two years of depression, began in 1983 to climb faster than on other European stock exchanges—it was losing the support of the workers, as measured in trade union elections. (France has four general labor confederations, divided on political and, in the past, religious lines, not counting confederations of managerial staff and of teachers. Their effective strength is best seen in polls at the factory level. But it can also be measured in elections on a national scale of delegates for labor courts (*prud'hommes*) or for joint bodies running social security. In the labor-court poll in 1982, the General Confederation of Labor (CGT), headed by the Communist Henri Krasucki, lost votes, while in the social-security poll in 1983, both the CGT and the French Democratic Labor Confederation (CFDT), the other union most closely associated with the government, suffered heavy losses.)[10]

The ideological swing to the Right of the middle classes is a reflection of the dual failure of the labor movement. In 1968, in the wake of the revolt of the students, many teachers, researchers, technicians, and other members of the professional intelligentsia were dazzled by the prospect of a different society and the role they might be playing in the transformation. Most sixty-eighters, discouraged by foreign shocks and the absence of domestic prospects, went back to cultivate their own gardens, while some of their spokesmen, as we saw, found a comfortable niche in the media preaching the reverse of their former beliefs. In 1981, Mitterrand managed to convince the middle classes that the Left, putting an end to stagnation, would reduce unemployment and improve their living standards. The honeymoon did not last. The Right, in fact, did not reestablish its hegemony through some heroic effort of intellectual resurrection. It did it on second-hand ideas borrowed from abroad because there was no opposition, the Left having no longer anything to offer.

The bankruptcy applied to both the Old Left and the New Left, which the French had erroneously nicknamed the *deuxième gauche* ("Second Left"). The Old Left in France included the Communist party, the CGT, the CERES of Jean-Pierre Chevènement, and quite a number of Mitterrand's own followers. It put the emphasis on growth, was in favor of nuclear development, civilian and military, and was, on the whole nationalistic. It relied on the extension of the public sector and on planning for the transformation of society from above rather than from below. Its failure was practical. The Left came to office and did not deliver: The growth was far from Japanese. Ideologically, the trouble with the Left was that, prisoner of its own coalition, it could not attack official policy frontally. It could criticize some aspects of the government's performance, not the basic promises of Mitterrand's line. Its arguments, therefore, did not sound credible.

For the Second Left, the bankruptcy was a more fundamental phenomenon. Second rather than New, the adjective is kept here on purpose because the New Left has been dead in France for quite a time. There, as elsewhere, the concept goes back to the 1950s; but the post-1968 New Left can be broadly identified with the May movement, with the forces that wanted to push the experiment as far as it would go, when Communists and Socialists alike were putting on the brakes. The Left so loosely defined was antihierarchical, egalitarian, and mistrustful of the central state. It talked of hegemony, of seizing power at all levels and not just conquering the Winter Palace. It was naturally open to the new social movements—feminism, ecology, the struggle against the nuclear invasion. This New Left died together with the May movement some time between March 4, 1972—the day of the funeral of Pierre Overney, the Maoist workman killed at Renault, its last mass demonstration in the streets—and the beginning of 1974—the end of the strike at Lip, the watch factory run by the workers and the symbolic experiment in self-management. There remained the *deuxième gauche*.

The relationship between the New Left and the Second

Left may be described in the terms of the television advertisement for a pseudo-alcoholic drink: "It looks like one, it tastes like one, but it isn't." The language sounded the same, and the names often were; but there was a basic difference. The New Left, aware that its projects had no chance of survival in the existing society, saw them as a weapon for society's destruction. The bowdlerized successor saw its reforms as props for the survival of an improved capitalist society.

Logically, Rocard should have been the spokesman of the Second Left after the Socialist victory. His career is a model. He had a radical past. As leader of the small but influential Unified Socialist party (PSU), he played a part in the May 1968 events and the year after was an eloquent candidate of the New Left in the presidential election. But five years later, he was rallying Mitterrand's Socialist party, although not with as many followers as he had hoped. The rest is well known. Having unsuccessfully challenged Mitterrand, he was compelled after 1981 to work his passage back, discreetly executing his ministerial duties. Jacques Delors, although never a radical like Rocard, might have also acted as the mouthpiece for the Second Left, but he was too busy with the economy. Thus the most articulate representative of this current was probably a third man, Edmond Maire, a chemical research worker by training and, since 1971, the leader of the CFDT. Under his predecessor, the CFDT had responded to the students in 1968 and turned the Communist-dominated CGT on its left. During Maire's leadership, the federation continued to talk of *autogestion* and experiment with new ideas, notably under the influence of the Italian labor movement, which is more sophisticated and original than the French. But as the social movement weakened and the crisis grew, Maire visibly mellowed and so did his union's policy. Still, when the Left came into office, the CFDT seemed to have a strategy, a conception of its own. It was ready to moderate quantitative demands, mainly on wages, in exchange for qualitative concessions over working hours, powers for the unions, and the general struggle against unemployment. Maire became angry with Mitterrand when the president decided that the one-hour reduction in the work week would involve no wage cuts for anybody. The CFDT, keen to cut the work week sharply in order to "spread" employment, was ready to limit full compensation to those earning, say, double the minimum wage. The union wanted to appear both radical and reasonable. But as the Socialist government renounced any further interference with the structure of capitalist society and the CFDT was an enthusiastic supporter of this moderation, Maire, forgetting his radicalism, became only too reasonable.

Such an epithet could well be applied to the Second Left as a whole. To talk about its bankruptcy was rather inaccurate. The year 1983 was, in a sense, its moment of triumph. Rocard, although condemned to modesty, was able to whisper under his breath: "I told you so." Mitterrand was virtually accepting the line of his defeated rival. But thus victorious, the *deuxième gauche* had lost its pink disguise. It was one thing to criticize the overextended state sector allegedly in the name of *autogestion,* of workers' control from below, and quite another to do so with the profit of private enterprise as the only alternative. It was splendidly progressive to attack the welfare state as overwhelming, distant, and bureaucratic. To do so without proposing new forms of social control by the associated users meant in our Western society control by the purse and, say, an even greater separation into medicine for the rich and medicine for the poor. No wonder that during Mitterrand's reign, the frontier almost vanished between the ideas of the Second Left, which liked to be called libertarian, and those of the Right "liberal" in the French sense of the term—the new champions of deregulation.

Maire illustrated this point in caricatural fashion toward the end of 1984 during the negotiations over the so-called flexibility. In practical terms, the employers were asking, in exchange for a vague promise of more jobs, for a removal of legal limits on their right to fire (they hitherto required the prior permission of a labor inspector) and on their right to determine the work week and year (to have, for instance, one longer week without overtime compensated by a shorter week). It is true that some political thinkers linked with the CFDT had argued in favor of labor à la carte, with a very flexible timetable. They had been writing, however, in a totally different context about the working people gaining mastery over their time, about a twenty-hour work week as part of a complete reshaping of French labor and leisure. In this case, even the delegates of Force Ouvrière, a trade union unsuspected of radicalism, could see that they were being asked to barter serious guarantees for peanuts and vague promises. The union leaders ultimately had to reject the deal under pressure from the activists. In the meantime, Maire the radical found himself on the right of André Bergeron, the head of Force Ouvrière, the favored partner of all governments and undisputed champion of class collaboration. It was revealing because not entirely accidental.

Ideological developments in Mitterrand's France throw new light on the old Socialist dilemma. By definition, the Socialists are carrying on their struggle within the framework of existing society, whereas the solutions they offer must lie outside the confines of capitalism. The risks are thus dual. If the Socialists are utopian and concentrate mainly on the project, they will find themselves miles ahead of their followers, the party becoming a sect. Yet if they limit themselves to defensive action, to the protection of the immediate interests of their members within the established society, the party will become its prisoner and, after a time, its upholder. The French Socialists in office, whatever their conservative critics may have said, were never threatened with the first danger. On the contrary, their experience shows the extent to which Socialist proposals, big and small, are ultimately linked with the global project, the vision of a different society. Taken out of this context, they do not make much sense.

How can one prevent the nationalization of industry from becoming an extension of state capitalism unless one tackles at one and the same time the problems of hierarchy, of the division of labor, of the control of the working people over their own work, and of the problem of power within society at large? To contrast the vicious state with the virtuous civil society without tackling the contradictions within that society is just empty talk. Or to take more concrete examples: Is it possible to take private property and the market not as facts of present-day life but as eternal metaphysical entities and project into this commercial society a pure radio and television in no way affected by their environment? Can one preach the virtues of commodity production and refuse to treat the press as a commodity? Should one view the school as an instrument for equality unless one has an egalitarian vision of society? If you agree that the ideological confrontation should take place on the terrain of your opponent and you also accept his premises, you should not be astonished if his logic does prevail. With this in mind, we may now review in quick succession the Socialist defeats, or should one say deviations, in the battles over radio and television, the press and the schools.

The radio and television, although linked, are really two separate stories. When the Socialists came into office, the state still had a full monopoly of both. There was nevertheless a difference. The radio knew no frontiers, as Pétain had experienced during the Second World War with people listening to the French service of the BBC. Besides, in addition to state-owned public radio, France had big commercial stations—Europe No. One, Luxembourg Radio, and Radio Monte Carlo—known as "peripheral" because their transmitters lie just outside French frontiers. In terms of political control, these stations did not present serious problems to the French government, which actually was a major shareholder in two of them. The new threat to the monopoly was technological. Radio transmitters for VHF were very cheap, and their expansion, sooner or later, was bound to affect French territory. In fact, some "pirate" programs had been broadcast to test the official reaction to the breach of the law, and the stations gained the flattering nickname of "free radios" (*radios libres*). The Right, as long as it was the beneficiary of the monopoly, responded to such tests by sending police officers to close down transmitters of the lawbreakers, which did not prevent it from emerging as the fighter for the freedom of the waves as soon as it was out of office.

Socialists were, on the whole, sympathetic to the free radios, and their victory was followed by a proliferation of these stations. A commission had to be set up to avoid total cacophony by limiting the power of the transmitters, allocating wavelengths, and encouraging various stations to merge. At the end of the process, there were some 60 authorized *radios libres* in Paris alone and close to 1,000 altogether. They were mostly poor, associative, community stations, relying on voluntary workers and small state subsidies. Was a great opportunity of inventing a different form of broadcasting missed here? It is doubtful whether such alternative stations could have flourished without a real development of democracy at the grass roots, and, in the best of cases, the massive extension of decentralized, public-sponsored broadcasting would have raised complicated issues of criteria and control. This is not the place to start a vast debate about how much can be started and achieved without altering the broader social framework. All I wanted to suggest, and which the alleged libertarians often seem to forget, is that in Western societies, the only alternative to the power of the state, with all its defects, appears to be the naked power of money. Freedom is coupled with commercial. The fate of the *radios libres* was sealed when the Socialists definitely endorsed the economic gospel of our system: Unto everyone that hath shall be given.

The inevitable happened. People with money invested in some of the stations on the assumption that advertising can be concealed and would, in any case, be allowed at some stage. Indeed, in his April 1984 press conference, on his return from the United States, Mitterrand confirmed that the law would soon fit facts rather than fiction. State subsidies would be granted to stations remaining noncommercial, while others would be legally allowed to take advertising. The limits on the money thus earned were set not so much because of Socialist principles, but because of the influence and electoral clout of the French provincial press reluctant to face too much competition for limited advertising funds. Even so, rich stations began to swallow the poor ones; broadcasting networks, although unauthorized, were established; and powerful transmitters were set up, breaking the law. What began as a social experiment thus had a commercial ending. Socialists can claim to have extended the freedom of the airwaves; only they did so *à l'américaine.* If there is still a big difference with the model, this is due to official controls and the survival of a large state-sponsored public service, Radio France, with its general broadcasting and its special cultural and music stations catering for minority audiences.

In television, state monopoly was complete in 1981: The three existing channels were state-owned. General de Gaulle, whose reign had coincided with the spread of television in France, claimed, wrongly, that the press was against him and that television, therefore, must be his weapon, its master's voice. President Georges Pompidou openly stated that television journalists must be aware that they are "the voice of France"; it sounded like the Voice of America and was meant in the same sense as an instrument of propaganda. Many journalists were kicked out after the 1968 strike, and even Gaullists were removed from key positions—"put in the cupboard" was the fashionable expression—when Giscard d'Estaing took over the presidency. The Socialists, coming after so many years out of office, were bound to reshuffle some heads of television channels and were urged by their supporters to remove some disliked faces from the screen. Taking into account the lag that the Left had in this domain, it was, if anything, less ruthless than its predecessors. Above all, it was the first to set up a screen between government and television, a body supposed to give the latter a degree of independence.

A law on radio and television, voted on July 29, 1982, provided for the creation of the High Authority, with nine members—three picked by the president of the Republic and as many by the leaders of the two chambers—renewable by one third every three years. It was to distribute permits for *radios libres,* appoint presidents of public radio and television channels, watch over political balance and fairness, and so on. The self-abnegation of the Socialists should not be exaggerated: The method of selection ensured them a provisional majority. But they did not pack the High Authority with stooges. Michèle Cotta, the chairwoman, an experienced journalist herself, was best described as a liberal, barely to the left of center. If the authority did fail on occasion—it was unable to prevent the government from appointing the man it chose as head of the second channel—television journalists were screened from direct interference and acquired new habits of independence.

The French Left was not breaking any new ground. The government never contemplated inventing a socialist television or using broadcasting as an instrument of social transformation.[11] France was simply catching up with neighboring European countries, Britain and the BBC acting as a model, where broadcasting may be accused of serving the Establishment, not of being a direct weapon in the hands of the government. Had the Socialists left it at that, theirs would not have been a negligible achievement. Mitterrand, however, had larger ambitions. Since freedom was the fashion and the United States, with its innumerable channels, the example, he was going to impress the French people by widening their choice. The project began as cultural and educational, only to change its nature on the way and become commercial; this, incidentally, was to be the usual course. Mitterrand entrusted the matter to his friend and *directeur de cabinet* André Rousselet, whom he appointed head of Havas, the big state-controlled advertising agency, and who concocted Canal Plus a, channel of paid television. Hardly a socialist innovation, it became a commercial success by the spring of 1986, when it reached 1 million subscribers for its diet of films, including an occasional hard porn late at night, and sports.

With satellites in orbit, television would soon, like radio, know no frontiers. Mitterrand was perturbed by the prospect of unfriendly conservatives, like, say, Rupert Murdoch, invading French screens. He had more immediate preoccupations, too, with the likely victory of the opposition in the parliamentary elections of March 1986. Once in office, the Right was likely to take control of a couple of channels and sell the other to political friends, leaving the president in not-so-splendid isolation. The preventive step was to set up a sympathetic commercial station. A lot of money was needed, and since wealthy investors were not flocking to Socialist headquarters, Mitterrand turned once again to the Schlumberger establishment, this time to Jérôme Seydoux, the grandson of the founder, and to Christophe Riboud, the son of Jean.[12] Funds had to be coupled with know-how, and Mitterrand remembered an operator, presented to him by Bettino Craxi, Italy's Socialist prime minister. He was *sua emittenza* Silvio Berlusconi, who had gained his nickname swallowing the bulk of Italy's private television and, say the critics, destroying its film industry in the process. Since the new channel, the Five (*la Cinq*) had to be ready by the end of 1985, before the election, its founders got the concession at a bargain price: special treatment for the first three years, with no guaranteed minimum for French production, and particularly favorable advertising terms, with the right for the first time in France to slice plays and films like sausages in order to fill them with commercials. When the agreement, duly signed, was revealed in November, there was an outburst of indignation and not only in the ranks of the opposition.

Mitterrand decided to go in and bat himself. That month, he devoted a great deal of the presidential press conference, a more ceremonial, roughly annual affair in France, to the defense of his project and a counterattack on the opposition. With Mitterrand's talent for irony, the Right was easy prey, as it landed itself in a ridiculous position. It found a new vocation as vulture for culture with a sudden love for public service and a contempt for "spaghetti television," when everybody knew its passion had been aroused by the fear that there might not be enough advertising for another commercial channel destined for their own merchants of tripe. The hypocrisy of your opponents does not ensure your own moral stature, and many Socialists were shattered by the move: Most affected, although not consulted, was the popular minister of culture, Jack Lang, who had begun his ministerial career sounding the tocsin against American "cultural imperialism." True, this was carefully misinterpreted, as though he had confused William Faulkner, or Louis Armstrong and his Hot Five for that matter, with mass-produced trash. Lang, under all sorts of pressure, had to pour a lot of cola into his wine. Still, he managed to be quite successful. Now, in the name of governmental responsibility, he had to keep silent and therefore approve, as France was introducing on its little screen the Berlusconi television—the European version of American commercial culture at its worst. It is debatable whether the mixture of old American serials and new Italian games gained any votes for Mitterrand. It is certain that the Socialists lost in this affair the moral right to lecture on the subject.[13]

The handling of the press legislation is another example

of what happens to a government that embarks on a venture without a compass and a clear sense of direction, although to understand the issue we must go back to its origins. On the morrow of the Second World War, the men emerging from the Resistance to the Nazis did seriously believe that they were going to live in another world. Among the things they hoped never to see again was a corrupt press, "instrument of commercial profit." General de Gaulle's postwar government codified some of the aspirations of the Resistance movement. Thus it seized all newspapers published beyond a given date during the German occupation; *Le Temps,* the mouthpiece of big business, died in this way, thus providing scope for the birth of *Le Monde.* It also passed an important decree, on August 26, 1944, making it compulsory for all publications to reveal their ownership and illegal for one person to own more than one newspaper. To pass such legislation in the heady days of the liberation was one thing; to apply it when society got back to "normal" was quite another.

At first, new capital was invested in weeklies and magazines, and then in provincial newspapers, consolidating their position through many local editions, rather than in the "nationals," those Paris papers read throughout France. (Before the war, national papers accounted for roughly two thirds of total French circulation; after the war, their share dropped to about one third.) After a time, however, the process of concentration resumed, and it was associated mainly with the name of Robert Hersant. Herr Sant is the way the satirical weekly *Le Canard enchaîné* spelled it for years to remind its readers that the man had begun his career by collaborating with the Nazis during the war. This did not prevent him afterward, starting with a car magazine, from building a newspaper empire. By the time the Socialists came into office, his group owned three Paris papers, accounting for over 33 percent of the "national" circulation, and fourteen regional papers, representing roughly 14 percent of the provincial press. His lawyers were busy running rings around the 1944 decree.

If the president had inspired the television offensive, the battle over the press was his prime minister's idea. Mitterrand, although the application of the 1944 decree figured among his electoral pledges, was not very keen on a confrontation concerning the press. His aide André Rousselet began negotiations with Hersant for the purchase of one of his two big dailies, *France-Soir*. Hersant, who then had troubles with tax inspectors, played for time and, finally, asked for impossible terms. Mauroy was thus allowed to proceed with the legislation as part of a drive to restore the prestige of the Left as a progressive force. By 1983, however, a Socialist government freshly converted to the virtues of private enterprise did not have the innocence and passion of the men of the Resistance to attack the corrupting power of money. It did not even propose some of the milder measures talked about within the profession, such as special privileges for newspapers run on a nonprofit basis or prerogatives for associations through which the editorial staff could get a share of decision-making power on a newspaper. It did not attack the monopoly exercised by some of the provincial papers in their regions. The new bill was merely a watered-down version of the 1944 decree, a mild antitrust law putting ceilings on ownership. Any one group was legally entitled to own either only provincial newspapers up to 15 percent of total provincial circulation, or both national and provincial papers, but in each case only up to 10 percent of the respective national and provincial circulation.

However soft the proposed legislation, it still would have

compelled Hersant to sell some of his newspapers, and this was much too much for the opposition. It fought the bill paragraph by paragraph with the zeal and vocabulary of freedom fighters resisting the power of tyranny. Nevertheless, the Left was bound to prevail in the National Assembly through sheer numbers, and in September 1984, the dreaded bill was finally voted. A month later, the Constitutional Council added the ultimate surprise. It found the limitations justified but their retrospective nature—illegal. The Hersant empire was thus safe. Socialists consoled themselves with the idea that it could no longer expand. It was an illusion. At the end of 1985, Hersant bought the big Lyons daily *Le Progrès,* establishing his monopoly in the region and mocking the new legislation. His friends would soon form a government, and he was laughing. Although the Mauroy mountain turned out to be a molehill, the whole episode contained a lesson. The Left was no longer able to build a moral case, not even against a lawbreaking reactionary press tycoon. The Right had stolen the flag of freedom, which it was now to wave triumphantly in the confrontation over Catholic schools.

The school battle was for the French people a sequel to the war of religion they had fought in the nineteenth century. The conflict in every French village and small town between the priest and the schoolteacher had been part of a vast confrontation between the Catholic monarchy, representing the feudal past, and the rising bourgeois Republic. It really began with the French Revolution, which just had the time to outline a treble framework—primary, secondary, higher—for public education. Napoleon codified the whole under the global title of University, proclaimed a state monopoly, but did little in practice about elementary education. In any case, the restoration of monarchy allowed the

Catholic Church to recover its prerogatives. Even when the "freedom to teach," the right to set up educational institutions, was firmly established by the Falloux law of 1850, the Church still preserved privileges. The balance was tipped in favor of the bourgeois Republic only with the introduction by Jules Ferry, through the laws of 1881 and 1882, of a vast system of "free, compulsory, and lay" education—*laïque*—that is, with no religion taught during the school hours. Free and compulsory applied only to primary schools, secondary and higher education being reserved essentially for the upper classes.[14] Many more battles against the alliance of "the sword and the holy-water sprinkler" were still to be fought, notably over the Dreyfus case, and the struggle was to reach its climax only in 1905, with the separation of the Church from the state.

Neither interests nor passions did disappear with the century. When the Third Republic collapsed during the Second World War, one of the first measures of Marshal Pétain's Vichy government was to ban the *écoles normales,* the lay teachers' training colleges, as places of perdition[15] and to introduce subsidies for Catholic schools. These were, naturally enough, abolished when France was liberated. Significantly, they were revived as soon as the Right recovered its position in the Fourth Republic, through the Barangé laws of 1951, and generalized when, the Left defeated, the Gaullists set up the Fifth Republic. The Debré law of 1959, passed despite mass protests by the advocates of lay schools, had one unexpected result. Up to then, the *laïques,* identified roughly with the Left, were claiming "public funds for public schools" alone. Afterward, they began to ask that all schools receiving public grants become an integral part of a "a great unified lay service of public education."

"Free schools" is simply French jargon for private schools,

and almost a synonym for Catholic ones, since all the others—Protestant, Jewish, lay—account for less than 7 percent of all the *écoles libres*. When the Left came into office, these essentially Catholic schools catered to just under 1 million pupils in elementary schools, close to 14 percent of the total, and just over 1 million in secondary education, accounting for over 20 percent of the total. The majority of these primary schools were run through a "simple contract," under which the state was paying just teachers' salaries; the bulk of secondary schools had a closer contract of "association," which bound them to respect the broad lines of the public-education program.

The *laïques* of the 1980s did not have the clarity of purpose of their nineteenth-century predecessors. Ferry and his colleagues considered that the bourgeois Republic had to offer the rudiments of a general education to its labor force if it was to develop. The Left coming into office under Mitterrand was not agreed on what a "Socialist" policy of education should be, and the slogan of a "unified, and lay service" was capable of many interpretations. Some argued that expanding and improving the existing system, with its principle of equal education for all, would progressively improve the chances of the unprivileged. Others replied that an alleged "equality of opportunity" in an unequal society was a sham. The 1960s, after all, had witnessed a heated debate, in which education was branded as, ultimately, an instrument for the preservation of privilege and for the perpetuation of the existing division of labor. Should a Socialist policy for schools abandon the imaginary "neutrality" and attack the very functions of the educational system? Should it develop the critical spirit of the pupils to prepare them to tackle the hierarchical division of labor? In short, could it be used as a weapon in the struggle for changing society?

All these questions, never clearly answered, were hidden beneath the cloak of a "unified and lay service."

Ideologically unarmed, the Left was also leaderless in this contest. Mitterrand, himself a product of Catholic education, had, as we saw, a very different background from that of his wife. The lay, to some extent anticlerical, tradition, a deep ingredient of the French Left, was alien to him. He paid lip service to the unified system, which figured among his pledges, but saw it as a nuisance and a potential vote loser. Actually, the way in which he treated the whole issue showed in a caricatural way how he conceived himself as "president of all the French." He did not act as a leader trying to carry through the interests of his side as the superior interests of society, but as somebody standing above the division and seeking a compromise. The politician he picked for the job of minister of education, Alain Savary, his unsuccessful rival ten years earlier for the leadership of the Socialist party, was undeniably a man of principle: He had interrupted his studies during the war to rally de Gaulle's Free French, and in 1957, he had resigned from Mollet's government to protest against the kidnapping in midair of Ahmed Ben Bella and other leaders of the Algerian insurrection. The task he was given was unenviable. He was to reconcile the Left, determined to integrate progressively all grant-aided schools into one system of public education, and the Catholic Church, equally resolved to continue receiving public money while ensuring the existence of a dual system.

The search for compromise rather than victory dictated a slow pace. Savary started his official consultations at the beginning of 1982 and produced a first version of his plan at the end of that year. This having been vetoed by the Catholic Establishment, it took him ten months and fifteen drafts to produce another. This time, the representatives of the Church did not reject the project outright, although they objected to many points, particularly the possibility of teachers in private schools becoming full civil servants and thus less dependent on their direct employers. (Strangely enough, teachers in public schools could think freely and, outside school, behave as they wished; in the so-called free schools, they could be sacked for being in favor of lay education, for divorcing, or for marrying a divorced person.)

A veto was now threatened by the other side. The Left felt that it had been cheated. True, some of the advantages of private schools were to be canceled. Thus new schools would be set up only in accordance with a national program; hitherto, the limitation applied to public schools alone. Such concessions, however, were small compensation for a total abdication on the central issue. The Socialist government, which was supposed to establish a "unified" system, was instead sanctifying its duality. At the very last moment, on May 22, 1984, when the debate began in the National Assembly, Mauroy, fearing a revolt in his own ranks, forced the hand of his minister and made a few concessions to the *laïques*. No money was to be given to set up private kindergartens in areas where there were no public ones. After a trial period of eight years, the local authorities would not be compelled to finance private elementary schools unless half of their teachers were public servants. These alterations, while not crucial, opened up possibilities of future change and were enough for a resigned Left to let the legislation get through its first reading. But the proposals provoked the wrath of the Church and of the opposition, both inside and outside parliament. This rejection was strange. After all, the risks were small and there was plenty of time to strike deals or introduce amendments, while the gain was immediate and once unthinkable: The Left was giving its official blessing to the public funding of private schools. If the Church now looked the gift horse in the mouth, it was because the balance of forces had shifted dramatically.

When the Right was in office, it had no democratic scruple on this subject. After the Debré Law on subsidies for "free schools" was passed in 1959, the National Committee of Lay Action collected 11 million signatures of adults to indicate that there was an electoral majority against it. The Debré government treated the petition with utter contempt as an insignificant scrap of paper. The Socialists refused to take advantage of their victory, thus inviting pressure to be put on them. To begin with, the pressure came from both sides. Catholics and *laïques* made a show of strength in Paris and in Brittany. Gradually, it became one-sided. The champions of the "free schools" developed the demonstration into a fine art. Large numbers of pupils were brought by their teachers. Members of the association of parents of Free Schools (UNAPEL) became efficient recruiting agents and masters of ceremony. Opposition parties lent a willing hand. An impressive structure was built. Altogether, the works, including even an anthem, the moving chorus of the Hebrew slaves from Verdi's *Nabucco*. Incensed, the advocates of lay schools staged one mass demonstration, on April 25, 1984, with over 1 million people throughout France, some 250,000 just in the capital. They were no longer a match, however, in a long contest. No wonder. They did not really know whether they were marching for or against their own government. The champions of "free schools" had no such problem. They had the Catholic hierarchy on their side—including Cardinal Lustiger, the archbishop of Paris—and all opposition politicians—from the center down to Le Pen. The demonstrations went crescendo: Bordeaux, Lyons, Rennes, Lille, and culminating in March with 500,000 marchers, symbolically in Versailles.[16] And after the Assembly vote, the clergy gave its blessing for the invasion of Paris on June 24.

Were they more or less than 1 million parading in the capital? It does not matter. The crowd was huge, and the demonstration was a tremendous success. The huge machinery had functioned perfectly. Over 5,000 parking places had been prepared for buses. Trains had brought people from all over France. All opposition parties had jumped on the bandwagon. When all this has been said, it does not explain how in a not deeply Catholic country, where the churches are empty, the defenders of private Catholic schools could present themselves as freedom fighters and successfully parade in such numbers. They were clearly filling an ideological vacuum for which the Left was largely responsible.

The distant nature of the bureaucratic apparatus, as we saw, facilitated attacks on the welfare state. In the school debate, the shadow of the state looked even more enormous. The Left might have gained a majority, although clearly not a consensus, for school reform as part of an egalitarian transformation of society. This was not at stake. Everybody knew that the young people emerging from this educational system would have to fit into a stratified society, that their social positions or even their jobs would be to some extent determined by their success. The Left was not just paying the price for the sorry state in which its predecessors had left public education. Nor was it just a revenge for 1968, the parents seeing in private schools a way of dominating their children. The spell was more complex. The denominational label of private schools should not conceal their class nature. They were essentially schools for the middle and upper classes; they took, for example, only a tiny proportion of immigrant children. But they also looked like a safety net, a solution of last resort for many more modest families that were very concerned with their children's education and faced with what seemed to be an alien Leviathan. The success of the Catholic Church was a reflection of the Left's inability to put political and social content into such concepts as freedom or equality.

Indeed, the school battle illustrates most graphically the journey of the Left toward ideological bankruptcy and recantation. First, it proposed an ill-defined educational reform linked with an even more vague project for social change. It then abdicated its role, entrusting the task to a leader who turned out to be a matchmaker. By the end of 1983, when the educational bill was finally drafted, the vision of a different society was abandoned altogether. The "free school" procession of June 24, 1984, thus marks the end of one act and the beginning of another. Mitterrand admitted his failure as a strategist and recovered his talents as a tactician. If you cannot beat them, join them. He would no longer pretend to be seeking a different solution. He would now stay openly on the opponent's territory, trying simply to improve his electoral chances. For this, he had to change his government. But to understand fully Mauroy's departure, we must first look at another defeat of the Left, in the European poll held on June 17, the Sunday preceding the triumphant Paris procession.

The European Economic Community has a parliament that is directly elected but has few real powers. The voters know it and often abstain. The election, staged every five years, is mainly significant as a large-scale opinion poll. In France, it is fairly accurate in one sense, since it is run through proportional representation on the national scale, and it is deceptive in another sense because voters, knowing that it does not matter much, may express their feelings without worrying about the consequences. Thus in 1984, discontented left-wing voters often stayed at home, bringing abstentions to the record level of 43 percent. Allowing for this, the verdict was still unmistakable. The Left suffered a severe defeat. The Communists were worse off. The attempt to stay in office and talk like the opposition proved a disaster. Their share of the vote dropped to 11.2 percent, back to the level of 1928. Their exit from government was now very likely. The Socialists dropped to 20.9 percent of the poll. They could expect to do better when the stakes would be more serious, but the hopes of a left-wing victory now looked very remote. The two parties of the respectable Right, running on a joint list headed by Veil, obtained a decent 42.7 percent. The real winner was Le Pen. His National Front, at 11.1 percent—almost level with the Communist party—captured ten seats in the European Parliament and the prospect of soon making an entry into the French National Assembly.

Within a week, Mitterrand suffered an electoral setback and a successful challenge in the streets. He required an immediate response. As the opposition had dared him to decide the school issue through a referendum and he had claimed that the constitution did not allow him to do so, he saw an opening. Let us stage a referendum, he suggested, to widen the scope of referendums in order to include issues connected with "public freedoms." The opposition, having asked for it, could hardly vote against the proposition, and it was difficult to present a man proposing an extension of democracy as a "liberticide." In announcing his new gambit, on July 12, the president revealed, incidentally, that the Savary bill on private schools would in any case be withdrawn. Having learned about the withdrawal on television, the author of the law wrote his letter of resignation at once. The prime minister followed suit, and the president would finally let him go after three years of devoted service. Once he had been a very popular lieutenant. He was now too worn to act as a shield. Something in this was quite symptomatic. When Pierre Mauroy took office, this cheerful social democrat from the north was viewed as a man on the right of the party; his appointment was interpreted as a safeguard against radical ventures. On the day of his departure, and although he had in the meantime presided over the policy of austerity, this man, still believing in the division of the country into two blocs and in the necessary alliance with the Communists, was perceived as very much on the left of his party. So long had been its journey. Socialism was out. Modernity was the new fashion. And the stage was set for Laurent Fabius, the young and smooth successor.

8

The Conversion

> I'd rather be an opportunist and float than go to the bottom with my principles around my neck.
> attributed to British Prime Minister Stanley Baldwin

Not quite thirty-eight years old, Laurent Fabius, smooth, handsome, and highly polished became in 1984 France's youngest prime minister since 1820.[1] Unkind critics likened him to Valéry Giscard d'Estaing, and not only because of premature signs of baldness. Like the former president, he was born with a silver spoon in his mouth, the son of a wealthy art merchant from an originally Jewish family that converted to Catholicism. He, too, had the ease and veneer of the upper classes, strengthened by the talent for passing exams. He was outstanding enough to be admitted to the Ecole Normale Supérieure, where he qualified in modern literature. He then entered the Ecole Nationale d'Administration, from which he graduated high enough on the list to join the State Council,[2] one of the top jobs available to civil servants. He was twenty-seven by then, and it was only the next year that the hard-working dandy discovered politics and the Socialist party. Another two years and in 1976, he was picked by François Mitterrand to be his "chief of staff," in practice his personal assistant. His career was thus linked to the fate of his master.

Fabius was at once given a constituency in the industrial area around Rouen, in Normandy, and soon after a seat in the National Assembly. He was Mitterrand's chief lieutenant, and a fighting one at that, in the battle against Michel Rocard. When Mitterrand became president, Fabius showed his paces first as junior minister in charge of the budget, and then as senior minister of industry and research. All this was grooming for the premiership. His performance revealed little about his predilections, since he was essentially his master's man. He had the skill, so appreciated among the *énarques,* of being able to plead anything and then its opposite. When the struggle against Rocard demanded, he was a preacher of socialism. At the Ministry of Finance, this child of a wealthy family was the author of the wealth tax (and if it excluded works of art, it was against his wishes, on specific orders from Mitterrand). Now, the demand being different, he was the champion of enterprise and efficiency. His slogan, on becoming prime minister, was "to modernize and to assemble." Mitterrand had picked the young and smooth technocrat to symbolize the new deal.

The break was the more obvious because the appointment of Fabius coincided with the departure of the Communists from the government. The link between the two events was not quite inevitable. The new premier did offer the Communists proportionately the same number of ministries. The Central Committee of the Communist party spent the night of July 18 in full session, while its leaders met with Fabius twice. The suspense was rather theatrical. The Communists were not bargaining over the number of portfolios; they were asking for a reversal of economic policy, which was just not on. The change of prime minister was a pretext. At nine o'clock in the morning on July 19, the decision was official. After three years in office, the Communists were leaving the government.

This is a good moment to look at the role they had played in it. The reason why we can tackle the issue so late is both simple and symptomatic. Since 1981, the Communists in government had had little influence on the shaping of policy. They carried their duties as well as their colleagues and argued as much about their own departments. Collectively, they had no genuine say in key decisions. Which does not mean that their previous presence in, or their future absence from, the government were irrelevant. The misfortunes of the Communist party under the reign of Georges Marchais are an important and intrinsic part of the vagaries of the French Left as whole.

In 1981, you may recall, a changed but unreformed Communist party was paying a delayed penalty for the break-up of the alliance. In April, the left-wing electorate had put Mitterrand well ahead of Marchais; in June, it aggravated its verdict, giving the Socialists more than twice as many votes in the parliamentary poll. When the two sides met, on June 23, to sign a common governmental platform, the Socialists could dictate the terms. The four Communist ministers were effectively political prisoners. It did not matter much for a year, since the government of Pierre Mauroy kept fulfilling its electoral promises. Trouble started in June 1982, when the Socialists embarked on a deflationary course, and became serious in March 1983, when they openly opted for austerity. As wages began to trail behind prices, as the indexing of the former on the latter was dropped, and, above all, as the "restructuring" of the economy affected the party's industrial bastions, the party was bound to react. For a while, it tried to march with the protesters and vote with the government.

On April 13, 1984, Marchais joined the steelworkers from the Lorraine who came to Paris to demonstrate against the government's policy. On April 19, the Communist deputies unanimously voted the confidence for that same government. If the Communist party expected to be rewarded for its political schizophrenia, it was mistaken. In the European poll, its electorate slumped disastrously by another third.

Critics, whether Socialist or Communist, were hindered by the conventions of Mitterrand's highly personalized methods of rule. The debate was open as long as Mitterrand's own position was ambiguous. As soon as he clearly sided with Mauroy, opting for classical austerity, to remain in the coalition, critics were allowed to question some aspects or consequences of the policy, not its basic premises. Thus they sounded neither coherent nor consistent. They were unable to outline a serious alternative without clashing with Mitterrand's capitalist solutions at home and his laissez-faire approach to foreign trade. And to claim that there was no excess labor in, say, the car or steel industries, as the Communists did, without at the same time constructing an alternative model, was the best way to lose credibility even among their own sympathizers.

Foreign as well as domestic events helped to tarnish the party's image. After the Second World War, the Communist party had profited from the prestige of the Red Army. Now it was the deserved victim of guilt by association. Before joining the government, it had approved the Soviet invasion of Afghanistan and, already within it, it gave its blessing to General Jaruzelski's coup. The Polish tragedy cost the Communists dearly and, incidentally, helped the broader campaign associating socialism with the stifling of freedom. If Marchais and his comrades had sided with the Polish workers against the military dictatorship, they would have been well placed even to question the strategy of Solidarity. If, cutting off the umbilical cord, they had provided a serious critical analysis of Soviet society, they would have been able to show the biased nature of much anti-Soviet propaganda, to stress the need of the capitalist Establishment for such a black foil.

The influence of the Communist party declined not only because of the shrinking of its constituency. Communists were no longer trusted even when they were in a mood of self-criticism. The exit from the government was followed by preparations for new assizes, and at the Twenty-fifth Party Congress, held in the Paris suburb of St. Ouen in February 1985, Marchais delivered a report that, in its own way, was a devastating indictment of the Communist conduct during the previous quarter of a century. The "collective intellectual," as Antonio Gramsci had called the party, had acted as a collective blunderer. Big business, Marchais admitted, had taken the structural changes in French society "in its stride much more rapidly than we did." Nor was it surprising, since "the thinking of the French Communists was deeply influenced by the model of society built in the Soviet Union." The Communist party had followed a defensive popular-front line, when a socialist alternative was required. It had wrongly "polarized hopes around a limited end and an electoral alliance." It had failed to perceive the potential of the upheaval of May 1968. It had spread the feeling that "the solutions to all problems will come from above," the illusion "that the transfer of property to the state will be enough to change it all." Missing all new aspirations, it had unwittingly paved the way for the Socialist monarch.[3]

Grasping past mistakes is usually half the way to a cure. Not in this case. The Communist party managed to admit past mistakes and preserve the principle of its own infallibility. Its methods of work were clearly beyond reproach, since there was no question of interfering with "democratic centralism," and the leadership must have been blameless, since the only members to be purged were the inner critics, such as Pierre Juquin, known as the *rénovateurs.* How could Communist professions of democratic faith be believed by outsiders when the party was doing simultaneously its utmost to limit the debate and to isolate and then purge the would-be reformers. How could the party's alleged conversion to *autogestion* be taken seriously, if in the months following the congress it did its best to show that in running its own affairs it could not conceive of a system of rule other than that of command from above.

Acting with the government and talking like the opposition, associated in the public mind with the tanks of General Jaruzelski, unable to provide a critique of Socialist policy based on a coherent, radical alternative, and belying its professions of democratic faith by its inner behavior, the party of Marchais was an easy target and a most convenient bogey. Yet it was not condemned mainly for its sins, which often served as a useful pretext. The violence of the anti-Communist campaign was dictated by the party's own radical reputation, by the roots it still had in the working class, by the strange connection between the existence of a large Communist party and the French belief in historical breaks, in the possibility of altering life by political action. In 1968, *Les Temps modernes,* the periodical of Jean-Paul Sartre, had written, "We knew we could not make a revolution without the Communists; we now know we cannot make it with them." The belief underlying much New Left thinking was that there could be no resurrection, no advance of genuinely radical socialism in France, without the removal of the Communist party, at least such as it stood. History, however, proceeded differently. The collapse of the party did not open the way to a radical revival, quite the contrary. Although the party through its conduct had helped to discredit a radical alternative, the shrinking of its influence was, paradoxically, a sign of the weakening of that alternative, of the "normalization" of France, its resigned acceptance of the permanence of the capitalist system. Whether this "normalization" is temporary will remain to the end the question dominating this book. On the fate of the Communist party, it is easier to venture a forecast at once. The Socialist move to the right gave the Communist party a respite and even a chance to restore its fortunes, which it did not seize. Short of a political and cultural revolution, which would turn the party into something quite different from a machine carrying out orders from above, the odds are that its exit from government in July 1984 marked merely a pause on the party's road toward historical doom.

Fabius had more immediate preoccupations. Communists had to be replaced, and the government, therefore, made to look as leftist as possible. This was achieved by three moves. First, the return to office, in the awkward Ministry of Education, of the critic Jean-Pierre Chevènement; with its leader back as minister, the CERES faction could be expected "to shut its trap." Second, Huguette Bouchardeau, that gentle remnant of the New Left, was promoted from junior to full minister of the environment. Finally, the moderate Jacques Delors, his domestic ambitions thwarted—too demanding in 1983, he was not even considered for the premiership this time—sought consolation in Brussels as head of the European Commission, and his place at the Ministry of Finance was taken over by the once reputedly more socialist Pierre Bérégovoy. All this, however, was window dressing. Mitterrand, who was now even more in charge of overall policy, with his former personal assistant as prime minister, had no intention of veering back to the left, far from it.

Shortly before the change of government, there was much fuss in Paris over an interview granted by the president to *Libération,* in which he insisted that France was and would remain a "mixed economy."[4] It was easy for Mitterrand's advocates, insisting on his continuity, to point out that this was not a new term in his vocabulary. In its emphasis on the concept rather than on its content, this was a good example of a phoney controversy. The most revolutionary regime taking over in a country like France would have preserved a "mixed economy"—part privately, part publicly owned— for quite a time. The only relevant question concerned the proportion between the two and the inner dynamics: In which direction would that society be moving? On this score, the views of the French president and of the Socialists had altered dramatically.

During the years in opposition, for instance, Jacques Attali, Mitterrand's faithful, could declare: "Socialism is the end of the realm of commodities; that is to say, the suppression of the market, the social appropriation of all the means of production, the self-management of all organizations, the decentralization of political institutions, the abolition of the division of labor."[5] After four years in office, Michel Rocard did not hesitate to tell his fellow *énarques* that the Socialists still asking for the socialization of the means of production "are a vanishing species."[6] *Verba volent, scripta manent.* This is no longer quite true. Because almost everything is now being recorded, it is easy and rather cruel to remind *all* Socialists of their previous reincarnation. (It becomes a duty only when, struck by amnesia, they stress their faithfulness.) Not to overstate the case, we shall limit ourselves to the president and, allowing for political license, deal with the gist of his thinking rather than damning quotations.

Even before becoming president, Mitterrand never pretended to be a Marxist; he merely claimed that Marxism was an important ingredient of his party's and, therefore, of his own heritage. While not fond of class conflict, he treated it as a fact of life imposed by capital. More indulgent than most of his comrades toward social democracy, he reminded the French of the social progress achieved in Scandinavian countries. At the same time, he blamed Social Democrats for not attacking the capitalist system in its very foundations. The French model was neither Soviet nor Swedish. Planning, at least on paper, was supposed to dominate the market, and the "mixed economy" was perceived as a period of transition, however slow, toward a different society. The 1984 version was exactly the reverse. The public sector was there to help the capitalist market to function properly in a "mixed economy" destined to last forever. Not surprisingly, the president told journalists at a breakfast meeting in March 1984 that "we must praise the search for compromise in all situations,"[7] and although he was not very fond of the word *consensus,* Mitterrand became its ardent practitioner. In this quest for the consensus, the Left dropped not only its Socialist principles, but also its Keynesian tools.

First, however, it had to get rid of the divisive school issue. In proposing a referendum to widen the scope of such consultations, the president had once again shown his tactical skill. The opposition, which had been clamoring for a referendum on education, chose to contradict itself rather than provide Mitterrand with an easy popular vote, since nobody would have pleaded against the proposed extension of freedoms. It therefore blocked the project in the Senate, where it had a majority. Although thwarted, the president could no longer be presented as a menace to freedom. As to Alain Savary's late bill, it was just buried. Chevènement, the new minister of education, changed as little as possible in the government's relations with the Church over the school issue and concentrated on the improvement of the state sector. His message was one of populist common sense. There had been too much talk of reform, critical spirit, and so on. One must now remember that the primary task of a school is the "transmission of knowledge." Reviving hard work and discipline, quantitative expansion should enable eight in ten French pupils to reach the *baccalauréat*—the high-school diploma—by the end of the century. Elitism does not alter because it is baptized "republican," and much of the minister's conception could be interpreted as conservative, but it was successful. For the president, the patriotic and populist Chevènement had the great advantage of removing this awkward obstacle from the center of the political stage.

His private assistant turned prime minister was also making an auspicious beginning. Handsome, he was made for the media. This product of sophisticated education was explaining things very simply on television with a studiously limited vocabulary, and since his socialist professions were limited to justice and solidarity, the message—notably a monthly quarter of an hour on television—did not hurt anybody. The ratings of Fabius in opinion polls were high; unfortunately, they did not drag along those of the Socialist party or of the president. With the parliamentary election only a year and a half away, time was running short for Mitterrand to shift his strategy. Unable to go down as the president of a Socialist new deal, he had to appear as the man who showed that the Left could last in office by managing the capitalist society more skillfully than its predecessor. For this, the dual task of the government was to consolidate the consensus, while balancing the books. Should this prove insufficient to restore the electoral chances of the Left, it would be necessary to alter the electoral law in order to prevent a landslide and thus leave the president of the Republic in not too uncomfortable a position.

If Fabius failed to achieve a consensus, it was not for want of trying. Entrusted with the management of France's capitalist economy, the Socialists decided to bring it up to date. Bérégovoy, at the Ministry of Finance, presided over the modernization of money markets, which helped to perpetuate the boom of the Bourse. Michel Delebarre, the young minister of labor, was not quite as successful, we saw, in fulfilling the other wish of big business—its demand for "flexibility," or the elimination of legal controls over employers' use of labor. The purpose, as Attali put it in private, was "to leave for the opposition not a bone to pick, nothing with which it could oppose us."[8] The Right would one day be grateful for the services rendered.[9] For the time being, concessions merely whetted its appetite. As the Socialists moved to the center, the conservatives swung far to the right. The year 1984 probably marked the climax of the French cult of Ronald Reagan. American expansion and creation of jobs were being contrasted with French stagnation and growing unemployment. French admirers echoed Reagan's preaching without bothering about his actual practice. When the Socialists in Paris started talking neutrally about nationalization, the opposition replied by proposing a program of privatization of the whole public sector, including the industries and banks nationalized by de Gaulle. When the government began loosening controls, the Right was vociferously clamoring for total deregulation.

Mitterrand's France was still the odd man out of the Western economy, but by now in reverse. It was tightening screws, while other countries were relaxing them. Austerity was slowly bringing its inevitable results. After two years of reduction of real wages and a cut in domestic consumption, the rate of inflation dropped to 4.7 percent in 1985 (with 2.5 percent a realistic target for the following year), while the foreign-payments deficit, which had reached 2.2 percent of the gross national product in 1982, was totally eliminated three years later. Even fortune seemed to be turning. The high rate for the dollar and the stiff price for oil having rendered the early years of Socialist management more difficult, their joint fall was brightening the prospects at the end of the reign. All this, however, could not balance the basic failure on the labor front, conceal the lengthening line of the jobless. Unable to provide an economic cure, the government invented a "social treatment" for unemployment. It involved subsidies to facilitate early retirement and, in a country where one in four persons under the age of twenty-five was jobless, the establishment of youth training schemes. The Fabius contribution was Collective Utility Work (TUC), under which young people were getting part-time work outside industry: half the minimum wage for twenty hours a week. This was better than nothing and improved the unemployment figures. With some obvious tricks, such as the removal of certain categories of jobless from the waiting lists for employment, the curve of unemployment was actually made to dip in 1985 to 10.1 percent of the labor force.

Few people were fooled by such statistical "miracles." The Left had not been elected to widen profit margins and boost the Bourse. Since it had failed to fulfill its own mandate, it was likely to pay a price in the next electoral test. Mitterrand, with his long experience and his feeling for the mood, sensed it in advance. The local, cantonal elections of March 1985 confirmed the finding of opinion polls: The

Left was barely above the 40 percent mark; the Right was approaching 60 percent. Even allowing for the midterm blues and the likely recovery of the government in the period preceding the election, the odds were against the Left exceeding 45 percent of the votes to be cast in March 1986. It was tactically important to switch to proportional representation.

Nothing requires quite the same mixture of high-sounding principles and down-to-earth calculation as a project of electoral reform. The Socialists were on solid moral ground. Proportional representation is relatively just. It may not always bring about strong government, but it gives a more accurate representation of political forces in a country than do majority systems. Besides, the Socialists had promised in their program to move in that direction. The calculation was even simpler. The system prevailing in France, like any other based on majority rule, amplified movements of the electorate. In 1981, the Left and particularly the Socialist party had been the beneficiaries of this bias. (If the French had voted as they had in 1981, but the seats had been won through the new proposed method of proportional representation, the Left, with 55 percent of the vote, would have gotten 57 instead of 68 percent of the seats, and the Socialist party, with its allies, 45 instead of 59 percent.)[10] The pendulum swinging in the opposite direction, it was bound to favor the Right in 1986. Proportional representation had not only the advantage of minimizing this victory. By giving a chance to the National Front to get a number of deputies, it might even prevent the respectable Right from gaining a majority on its own. And while you are at it, you might as well go whole hog. Amid the existing variants of proportional representation, Mitterrand opted for the one—the biggest average and not the biggest remnant, with calculations on the departmental and not on the national scale—that favored the bigger parties—the Socialists rather than the Communists or the smaller leftist parties. Indeed, to discourage people from voting for the Greens or the Far Left in the two very large departments, the North and the Paris region, where candidates could get elected with a relatively small proportion of the vote, it was decreed that those obtaining less than 5 percent of the votes cast would just not be elected to parliament.[11] The electoral law had been made to measure.

It nevertheless did claim an unexpected victim. The final decision on electoral law was taken by the Socialist government on April 3 in the morning. That very night, after 2:00 A.M., the French press agency reported Rocard's resignation. Two days later, *Le Monde* printed his principled objections to the law: France needed a stable executive, the deputies should not be chosen by the central party machine, and deals should be made publicly before the election and not secretly afterward. But there were also less theoretical reasons for his move. Condemned for his original sin of having challenged Mitterrand, Rocard had spent almost four years in governmental purgatory, serving silently in planning and then in agriculture. He was not even able to gloat openly as his party was rallying to his moderate positions. All this was of no avail. In the last reshuffle, he had been refused the Ministry of Finance, and there was no sign that Mitterrand would ever favor the presidential ambitions of his former rival. It had not mattered too much as long as Rocard had remained the undisputed darling of opinion polls. The rising star of Fabius was casting a shadow over his own performance. It was no longer possible to parade as the "modern" as contrasted with the "archaic," the man of the new generation as opposed to the worn-out veteran. The difference of age between Fabius and Rocard was greater than between him and Mitterrand.[12] For Rocard, the aspiring presidential candidate, it was now necessary to take a distance, to emphasize his own profile, and to let the prime minister take the blame for vicissitudes of office. In the summer following Rocard's departure, the Socialist government was going to suffer its only major scandal.

When a party, even a catch-all party, is reduced to the function of an electoral machine, when its militants have to eat their words and swallow past pronouncements, when the gap between promise and fulfillment grows too big and, since the change of line was never justified, so does the gap between current discourse and current performance, when people who talked of changing life must limit their ambitions to preserving some powers for the president, something then happens not just to the morale, but also to the morality of a party. In fiction, the corruption and betrayals may not be apparent because they distort the horrible portrait hidden in the attic. In the realm of politics, they cannot be concealed for long. The Greenpeace scandal, or rather its handling by the Socialists, was such an eye-opener, a revelation of the moral debasement of the party.

On July 10 at almost midnight in New Zealand's Auckland harbor, two bombs exploded in quick succession below the hull of a trawler, the *Rainbow Warrior,* sinking it and killing one of the crew, Fernando Pereira, a Dutch photographer of Portuguese origin. The boat belonged to Greenpeace, the international environmentalist movement, and was the flagship of a flotilla sailing to protest against the forthcoming French nuclear tests on the Polynesian islands in the South Pacific. Were it not for the nuclear background and the tragic death, what followed was farcical. With French secret agents telephoning Paris without precautions and leaving clues all over the place, it was like a James Bond film with gadgets and all, but with Peter Sellers—alias Inspector Clouseau—acting the main part.

Two days after the explosion, the New Zealand police arrested a "Swiss couple," the Turenges, whose forged passports did not stand up to international checks. The "wife" was Dominique Prieur, a captain in the General Directorate for External Security (DGSE, the French equivalent of the CIA). The "husband" was Alain Mafart, a major in the same service and former deputy head of a training center for frogmen that the agency had at Aspretto, in Corsica. Actually, French people had been swarming all over the place. A young Frenchman had visited the boat on the day of the explosion. A woman, allegedly a "sympathizer" of Greenpeace, in fact another captain of the DGSE, had spent quite a time in Auckland collecting background information. In June, four men, three of them petty officers from the Corsican center for frogmen, had arrived by boat from New Caledonia; their yacht, the *Ouvéa,* subsequently vanished in mid-ocean. Many more agents were still to be discovered.

Meanwhile, Paris reacted as though it were not concerned. The minister of defense, Charles Hernu, an old friend of Mitterrand in love with the armed forces, assured the government that his services, including the DGSE, were in no way involved. This led to some comic misunderstandings, as the French police, acting on queries from New Zealand, were discovering more and more proofs, while the military, from the minister downward, were denying with abandon. It was not much use. The prosecutor's finger was pointing inexorably toward Paris. On August 6, the president summoned Fabius to see him. It was high time to act, since the Paris papers were about to reveal the French connection, but the authorities thought they had a clever counter. They promised to punish the guilty, whatever their rank, and set up an independent inquiry headed by a member of the opposition, Bernard Tricot, former chief of staff of General de Gaulle. The assumption that as a Gaullist he would opt for "national grandeur" and as a civil servant would know the art of covering up was wise. In his report, published on August 25, Tricot recorded and endorsed all French protestations of innocence. "Too transparent to be a whitewash," summed up wittily David Lange, the prime minister of New Zealand. Tricot himself seems to have had afterthoughts, saying on television the day after publication: "I do not exclude that I was led astray." As to Fabius, he deliberately distanced himself from this gift horse—"At this stage, I have no element to lead me to counter this conviction," referring to Tricot's presumption of French innocence—knowing well that the preliminary hearing in Auckland, scheduled for November 4, might blow the whole Tricot report to pieces.

He did not have to wait so long. On September 17, showing that it could also shine in investigative journalism, *Le Monde* splashed on its front page the story of a "third team," which had actually stuck the bombs. The reporters[13] did not furnish definite proofs, but their carefully researched story put the various pieces of the puzzle together and sounded very convincing. Hernu, in keeping with Napoleon's advice, tried to defend himself by attacking the press and all those who dared to suggest that the military might not be a paragon of republican virtue or, horrible thought, might occasionally depart from the truth. That a "Socialist" minister should utter such nonsense in a country where the Dreyfus case had rested on plain forgeries perpetrated by the general staff and where the Fourth Republic had collapsed under the blows of military commanders was, to put it mildly, surrealist. This, however, was Hernu's last ammunition. The prime minister had grasped that he must act quickly or he would sink together with the *Rainbow Warrior,* and, under pressure, a reluctant Mitterrand had to accept the departure of his old companion. Hernu, incidentally, went with full honors, the president greeting his resignation, on September 20, with "pain, sorrow, and gratitude." The only man to go with him was Admiral Pierre Lacoste, the head of the DGSE, who had obtained guarantees that none of his subordinates involved in the case would be punished. Two days later, Paul Quilès, the new minister of defense, had all the information, and Fabius could confess on television that French agents had sunk the ship and had acted on orders, and we did not know. A tough general was appointed as the new boss of the DGSE determined to keep things quiet. The "swimming pool" (*la piscine*), as the headquarters of the French secret services are called, was no longer supposed to leak. Domestically, it was the end of the affair.

There still remained the couple in distant New Zealand. They were now in a position to alter their plea to guilty in exchange for a switch in the indictment from murder to manslaughter. French papers—and ministers, for that matter—started describing the case as though it were settled, with the "heroes" bound to get home by Christmas. They were in for a shock. On November 22, the verdict was announced, and it was stiff: ten years in jail. It took months of commercial blackmail, particularly over EEC imports of New Zealand butter, and a change of government in France for a deal to be struck, with Javier Perez de Cuellar, secretary general of the United Nations, acting as matchmaker. (France agreed to send a letter of apology, pay $7 million in damages, and lift its veto on New Zealand imports to the Common Market.) In July 1986, a year after their arrest, the "Turenges" flew to Hao, a French military base some 560 miles east of Tahiti, where they were supposed to stay for three years out of the limelight.

Seen in perspective, the whole operation looks not just criminal, but absurd. Even if it had been "successful," meaning nobody hurt and nobody caught, the sinking of a Greenpeace vessel sailing for a protest against French nuclear tests would have been branded by everybody as French handiwork and would have produced more adverse publicity than skirmishes with pacifists did on previous occasions. This was not the opinion of the admiral in charge of French nuclear tests, who allegedly was worried by the size of the *Rainbow Warrior* and early in March 1985 had asked the Ministry of Defense to do something about it. Hernu, the son of a gendarme who as a child must have dreamed of medals and uniforms, saw here an opportunity to play cloak-and-dagger games. Was the president in the picture? In broad terms, he must have been, since the operation, if only for financial reasons, needed the approval of his personal chief of staff, General Jean Saulnier, who, far from being downgraded, was promoted chief of general staff of the French armed forces. The president may well have been told by Admiral Lacoste that "there will be no victim, no trace, and France will not be implicated."[14] But the president's prior knowledge is not required to explain his covering up for Hernu. Mitterrand, who spent twenty-three years in the wilderness, is not a man to forgive or forget, and Hernu was one of his rare companions throughout the period. The man who appears as an outsider in this story, trying all the time to get on top so as not to be swept aside, is Fabius, and this is the first sign that the sympathy between the president and his favorite son may not be quite perfect.

Why so much indignation? Cannot medium-size states get away with murder, or is the perpetration of crimes in the name of the alleged national interest the privilege of superpowers? Is not the DGSE entitled to blunder, considering that the CIA cowboys have done it so often? The answer to such objections is that the French wanted all this and the mantle of socialist morality too. The *Rainbow Warrior* episode was a confirmation that in foreign as well as domestic affairs, the French government's only link with socialism by that time was its name. The scandal never threatened to become a French Watergate because both sides were tied in jingoist unity. In the midst of the crisis, President Mitterrand went out of his way to proclaim that the nuclear tests would go on as long as "France deemed it necessary" and urged French troops to repel any interference. The opposition dared not criticize too much so as not to be branded as "unpatriotic." There were few voices in France to condemn the attack on the pacifists and not its clumsiness. When talking with Socialists at the time, one always had the impression that at the back of their minds was the thought: If only that bloody boat had been blown up in mid-ocean. How far the Socialist party did sink was shown in October in Toulouse, where at the party congress the biggest ovation, the martyr's and hero's welcome, was reserved for none other than Charles Hernu. (To stop here would be to present the story in Manichean terms. The second biggest ovation was for Robert Badinter, and the congress was also greatly stirred by greetings for Nelson Mandela. The ways of man, too, are sometimes mysterious.)

Toulouse, known as the pink city because of the color of its brick walls, was in October the setting for a not very Red performance. The congress of the Socialist party, the last before the elections, was presented everywhere as the French

Bad Godesberg, by analogy with the meeting at which the German Social Democrats had made their formal break with Marxism. It was also expected to mark the triumph of the man who symbolized this evolution, Michel Rocard, the self-proclaimed "destroyer of dreams." It turned out to be a climax in the career of Laurent Fabius. The three strands were not unconnected.

French Socialists had shifted their position without acknowledging the move and justifying it in pragmatic terms of realism, inevitability, and, the new euphemism for opportunism, "the culture of government." It was conversion without agonizing reappraisal, apostasy without ideological fuss. Among leaders suffering from acute amnesia, to be called a social democrat, yesterday an insult, had become a name of praise. Yet they also knew that French traditions and the structure of French labor unions were not favorable for the growth of social democracy. The logical transition was toward a French equivalent of the American Democratic party,[15] and this, even allowing for the collapse of the Communist party and the disappearance of an alternative, was a tall order.

No longer bound, since April, by collective governmental responsibility, Rocard was able to go beyond whispering "I told you so." He urged his party, now that it had stopped parading as the gravedigger of capitalism, to recognize and analyze its past mistakes. The prospect of recantation did not delight his colleagues and, nevertheless, rivals. But this idea of candid admission was probably popular in the country, since Rocard preserved high ratings in opinion polls, and even within the ranks of the Socialist party, since in the voting that preceded the Toulouse congress the Rocard resolution had been approved by 28.6 percent of the rank and file, a few points more than expected. On the eve of election, conflict was to be avoided, and various views would be merged into one composite resolution. Rocard traveled to Toulouse to reach a compromise, certain this time to be the star of the show.

He was outwitted by Lionel Jospin, who in the four years since he had inherited the leadership of the party from Mitterrand had learned a great deal about its control. Jospin chose to outflank Rocard on his right. If we don't want to be a small party hovering to the left of center, he argued, we would not mind at all being a big social-democratic one like the German or the Swedish; only circumstances being different in France, we must invent our own solution. His political clothes stolen, Rocard did not know which posture to assume. He who occasionally could dazzle his audience, sent it to sleep with a muddled and fairly incomprehensible speech. To make matters worse, the limelight on the last day fell on a Fabius who was in splendid form, calling his fellow leaders rhetorically by first name—you, Pierre; you, Michel; you, Lionel—to joint struggle and who was perceived by the delegates to the congress, preoccupied with salvaging their position, as Mitterrand's first lieutenant, capable of leading them into electoral battle. The triumph was his.

Our modern star system is supposed to encourage meteoric careers, the mass media both making and breaking them. In the exaggerated version prevailing in Paris, the turning point in the career of Fabius, a product of television, occurred on the little screen. Two weeks after the Toulouse performance, on October 26 to be precise, Fabius was to confront on television the champion of the Right, Jacques Chirac, the neo-Gaullist leader and former prime minister. On your left Kid Fabius, on your right Battling Chirac; although promoted like a Las Vegas contest for the undisputed heavyweight championship, the match did not come up to expectations. Not because the protagonists pulled their punches. They did not. ("I have seen liars, but . . . ," exclaimed Fabius, while Chirac accused him of "barking like a pug-dog.") If anything, they often hit below the belt. But their apparent passion was phoney. It did not reflect a deep, ideological division; it concealed its absence. Consensus, paradoxically, was leading to fisticuffs. This was particularly true for Fabius, who, accused of not sounding socialist enough, tried to make up for ideological timidity through verbal violence. The smooth and gentle technocrat was suddenly perceived as aggressive and arrogant, which apparently did not do much good for his image. The impact, however, has been overstated. The picture of Fabius, the champion of modernity, had to be tarnished after a time. A prime minister is judged by the performance of his government, and sooner or later, Fabius's popularity ratings were bound to move toward the lower levels reached by his president and party.

Another element in the downgrading of Fabius was the almost inevitable strains in his romance with Mitterrand. At the beginning, he had to assert his very existence: "Him it's him, and me it's me," he proclaimed in one of his early appearances on television as prime minister. With time, he acquired stature, interests of his own, and, by the same token, risks of conflict. The first sign of tension emerged in June, when Fabius decided, well in advance, to launch and lead his side's electoral campaign. He opened it with a meeting in Marseilles and met with immediate resistance. Jospin threatened to resign if the Socialist party was forced to take a back seat in the campaign. To the surprise and sorrow of Fabius, Mitterrand refused to choose between his two favorite sons: He decreed, Solomon-like, that the government and the party should each run its own campaign. Then came the stresses of the Greenpeace scandal, during which Fabius, fighting for his own skin, forced the president to get rid of one of his old companions. Finally, in December, the perfect match almost ended in divorce. For reasons that are still not too clear, in a country where the Polish affair had had such a strong sentimental echo, the president decided to be the first important Western head of state to act as host to General Jaruzelski. Fabius was as shocked and almost as surprised as the average Socialist rank and file; the visit had been negotiated behind his back. Answering a question in parliament, he stated plainly that he was "personally troubled" by this visit, summed up objectively the president's reasons for staging it, and concluded with a brisk "I have nothing to add." In a republican monarchy, in which the prime minister was hitherto acting as a shield for the president, this was interpreted as a case of lese majesty. For several days, the commentators and apparently the two protagonists toyed with the idea of resignation or dismissal.

Inevitable dissensions in a ruling couple had, probably, little influence on the actual course of events. Mitterrand was, in any case, a man to have several irons in the fire, and since the general election was approaching, he was always likely to move up to mastermind the maneuvers. To cut the losses in order to remain fit to fight another battle had been the major art of his career, and this exercise had now become crucial. If the Socialists were swept aside by an electoral landslide, the man at the Elysée would not be able to resist on his own. If the result were more moderate, however, the president might be able to coexist with the opposition, to indulge in a period of *cohabitation,* the modish term used rather than "marriage of convenience" to describe this liaison, an expression that became fashionable as political attention focused on the two years separating the parliamentary poll of 1986 from the presidential election of 1988.

Thinking of his own future, Mitterrand was doing his best to improve his party's position. In addition to the advantages already mentioned, the new electoral law allowed him not to bother about the split with the Communist party, at least not in immediate electoral terms. The prospect of the presidential poll, in turn, revived rivalries on the right. Politically, these divisions were not as clear-cut as they had been in the past. In General de Gaulle's time, there were his followers, the Gaullists—authoritarian, relying on state intervention, populist while speaking of workers' "participation" and nationalist when opposed to "American hegemony"—and the more classical conservatives—favoring free trade at home as well as abroad and a Western alliance under American leadership. President Georges Pompidou, on whom the Gaullist mantle fell, blurred this separation. He represented the immediate interests of the bourgeoisie and not any long-term visions. Chirac, a pupil of Pompidou, carried this transition to the point of caricature. On the eve of the parliamentary election of 1986, he was against any state intervention in the economy, was in favor of privatizing the firms nationalized by de Gaulle, and criticized Mitterrand for not taking part in the American Strategic Defense Initiative (SDI) program, enough for the general to turn in his grave. The fading of frontiers did not result in a reduction of rivalry. Chirac was at the head of the biggest party of the Right, the Rally for the Republic (RPR), but his standing was low in opinion polls. Raymond Barre, whose smug satisfaction while forcing the French to tighten their belts had made him a most unpopular premier, had since had his fortunes restored by the Socialist volte-face. Although his ratings were now high, he had no party machine. Thus Chirac, as leader of the biggest conservative party, was likely to be picked by Mitterrand in case of a clear conservative victory, and he had every reason to coexist with him and build his reputation as prime minister. Everything, on the contrary, pushed Barre to condemn *cohabitation* as sinful. United against the Left, Chirac and Barre could not hide their bitter competition.

And so we are back to where this book began, back to the strangely dispassionate election of March 1986. Mitterrand had done a lot, exploiting the divisions of the Right and the resignation of the Left, to improve the chances of the Socialist party. He did not commit himself to follow it in its fall from power. Unlike de Gaulle, at no stage did he stake his own future on an election that did not concern him directly. He knew that the Left, not having delivered what it had promised, had no chance of winning. It was just a question of minimizing the impact of its defeat. The result came fairly close to expectations. The swing was hefty. The Right defeated the Left by roughly the same margin by which it had been beaten five years earlier, and under the circumstances, Mitterrand could not pick and choose: He had to call on Chirac to form a government. The verdict, however, was also sufficiently blurred for him to stay on. Indeed, since the Socialists emerged as the biggest party in a National Assembly in which the respectable Right on its own enjoyed only the narrowest majority, with his tactical skill and a bit of luck, Mitterrand even stood a chance of revenge in 1988.

Although Mitterrand carries on as president, it is a good point to stop our narrative. The Socialist experiment, had it ever been on, was now definitely over. The coexistence between a president elected on a left-wing ticket and a new conservative Assembly is quite a different story. Naturally, there is a link between the two periods. *Cohabitation* puts the previous five years into focus and is inconceivable without them. If nationalization, to take one example, were still considered to be an instrument of planning and social transformation, as it had seemed to be viewed by the Socialists before their governmental venture, Mitterrand could not have stayed in office presiding over its dismantlement. As things had evolved, he felt quite comfortable. In the twin fields of defense and foreign affairs, where the president has constitutional prerogatives, the differences between him and the opposition, as we shall see in Chapters 9 and 10, were really marginal. On the domestic front, where the government is supposed to govern, Mitterrand was able to intervene occasionally as a responsible middle-of-the-road guide, a keeper of the consensus protesting against ideological excesses and dangerous ventures. As students took to the streets and workers began to strike again, he could and did issue warnings about how unwise it was to apply economic austerity without social dressing, how foolish it was to cut taxes ostentatiously for the rich, while asking workers to tighten their belts. The president paraded as an elder statesman, a more sophisticated upholder and modernizer of the existing regime.

Had Mitterrand altered in the five years spent at the top? Physically not much. He was slightly thinner and stiffer, walking as if he had swallowed a stick. He who had always done his best to conceal his innermost thoughts ("He is so frightened that one may read his decisions on his face that he hesitates to entrust himself with the totality of his thought," quipped Jean Riboud)[16] now looked perfectly inscrutable. Although simple in his private life—fond of reading, writing, and watching trees grow—he appeared slightly pompous, dignified in his public appearances. True, he could star in a popular television program to show that he understood the latest slang, the *verlan*,[17] and the sly politician was never far below the surface; yet the dominant impression was of a self-conscious man terribly aware of the importance of his role and function.

Clearly, the question was not mainly concerned with appearance. Earlier in this study we portrayed Mitterrand at the opening of his reign as a tactician turned strategist, a politician aspiring to be a statesman, a man keen on leaving a mark, an imprint, and determined to change the course of history. Did he? Obviously not in the direction he had been expected to or had promised. In spite of nationalizations, France was farther from a socialist horizon at the end of Socialist rule than at the beginning. But here we must deal with the cynical objection: The message did not matter. At the time and in the place he was, he had to use such language to get a hearing and to outbid the Communists. Once victorious, he could discard it. There is something in it. But although anyone who had heard Mitterrand, the already veteran politician, suddenly switch to the language of class struggle at Epinay in 1971 will have to admit his ability to adapt the vocabulary to the audience; the purely cynical interpretation is much too simple.

Southerners, it has been said in France, do not lie; they believe in what they say. Yet what people keep on saying is not immaterial. Blaise Pascal advised men to kneel and pray, on the assumption that faith would follow, while modern psychology suggests that the act of crying has an influence on the mood. The discourse developed by Mitterrand over ten years, from Epinay to the presidential election, was not just make-believe. The logic of opposition, the influence of more radical companions, the general mood after

1968, and the fate of Salvador Allende drove this middle-class reformer to perceive himself as the leader who would somehow defy "the powers of money" and would lead exemplary France somewhere beyond social democracy. All this gradually, peacefully, within existing institutions.

The belief, admittedly, turned out to be skin-deep. When the means and methods employed revealed themselves to be insufficient to achieve the ends, when the domestic and foreign resistance called for more radical solutions, Mitterrand gave up the whole project altogether. With the skill of a genuine pragmatist, he switched purposes as others switch horses in midstream. Albert Camus once accused Jean-Paul Sartre, unfairly, of "turning his armchair in the direction of history." Mitterrand was thus turning his republican throne. The great task of our president, Mitterrand the modernizer, claim his panegyrists, is to bring the French Left into the mainstream, to show that it is capable of governing, to ensure that *alternance* becomes a permanent feature of French political life.

Alternance does not require a name in Britain or the United States, in Germany or Scandinavia. It is the simple fact that Labor follows Conservatives, Democrats replace Republicans, Christian Democrats succeed Social Democrats, and vice versa. The rule was the exception in France. The Left in office was treated as a rare visitor, an intruder: It stayed for barely a year as a popular front in 1936, and somewhat longer after the Second World War, but partly under the general's supervision. There was a reason for this difference. In the United States and in Britain, the battles, however bitter, never threaten the institutional framework. In France, the Left was suspected of not respecting the rules, of dreaming of another society. Mitterrand the "normalizer" is supposed to have put an end to this exceptional status of the French Left, to have rendered *alternance* permanent by shattering the prospect of a radical alternative.

Whether he should be praised or blamed for this normalization depends, naturally, on one's political outlook. But is this achievement really historical or merely transient? This is one of the questions for the conclusion. Before passing judgment, we must envisage another possibility—that the failure was inevitable, that if France had had a perfect Socialist government, if the movement were genuine and conditions favorable, the project would have failed nevertheless, because it is national and the nation-state no longer provides the scope for its fulfillment. Before moving to the conclusion, we must still clear one hurdle and examine Mitterrand's performance on the international stage.

III

The European Dimension

The Sleep of Reason Produces Monsters
Goya, etching

Compared with François Mitterrand, even the Polish pope may be described as sedentary. Trying to justify his peregrinations,[1] Mitterrand pointed out that quite a lot of international traveling is now compulsory routine for a French president. In a year, he must attend about three summit conferences of the European Economic Community, and meet with the German chancellor at least twice and the British and Italian prime ministers once or twice. This is a minimum, to which should be added conferences of the seven big industrial powers, the Franco-African summit, and so on. On top of it came the brief diplomatic encounters and the longer state visits, of which Mitterrand was visibly fond. He traveled to Mexico and Brazil at one end of the world, to India, China, and Japan at the other. He was the first French president to pay an official visit to Israel, but he also went to Cairo, Amman, and Damascus. He talked about Chad with Colonel Muammar Qaddafi on Crete and flew to

Beirut overnight when French soldiers were blown up there. Like General de Gaulle, he was highly preoccupied with France's standing in world affairs. His journeys, however, did not have the impact of the general's, nor did the results of his foreign policy come up to Socialist expectations.

The expectations had been quite high. Nobody thought that the Socialist president would be "softer" in his relations with Moscow than were his predecessors. On the contrary, Mitterrand had been tougher in condemning the invasion of Afghanistan and was likely to stick to his line in office. But although nothing in his political career suggested any neutralist temptations, he was expected to be equally harsh in his judgments of Washington's conduct in Central America. Altogether, cutting drastically the sales of French arms and inaugurating a new treatment of countries like South Africa, the Socialist presidency was to offset through a new moral standing its relative lack of power. French diplomacy would thus be in a position to carry out its main task—to shift the emphasis in international relations from "East–West" to "North–South," which in the jargon of the time meant putting the aid to poor countries as the first item on the agenda and radically altering economic relations between the industrial Northern Hemisphere and the "proletarian nations" of the Southern.

France, in this version, was to be the interpreter not only for the countries of its former African empire, but also for the have-not nations altogether. Some Socialists were even more sanguine. Mitterrand, they assumed, had an opportunity to fill de Gaulle's otherwise empty "grand design" with real social content. With nationalization and planning, France had the means to strive for genuine economic independence and, acting as an attractive model, might even initiate a movement toward a United States of Europe, alone able to question American hegemony. Yet this was daydreaming. In foreign as well as domestic policy, the gap proved tremendous between the promise and the reality. The unfulfillment in both cases is intimately connected.

9

Reagan's Best Ally

Americans who saw the new French president painted as a pink intruder with dark red forces hovering in the background did not have to wait for long—that is, until his belated domestic conversion to the capitalist gospel—to discover that there was something wrong with this image. It clashed with another, which was emerging rapidly and was showing François Mitterrand as Ronald Reagan's closest companion in the crucial confrontation with the Soviets, the battle of "Euromissiles." The members of NATO, you will recall, had agreed that the replacement by the Russians of their old intermediate-range missiles—very roughly, those capable of hitting Western Europe but not crossing the Atlantic—by the more modern and more accurate SS-20s justified, or could serve as a pretext for, the deployment of American Pershing-2 and Cruise missiles in Western Europe.[1] As the abstract decision became a concrete proposal, it not only provoked the Soviet threat to break all disarmament talks, but also revived the antinuclear movement in Western Europe on an unprecedented scale. Some politicians began to have cold feet. France, not a member of NATO's integrated military command, was not directly involved; there were to be no missiles on French soil. The whole-hearted support of this uncommitted partner, and a

Socialist one at that, was for Washington a godsend. Mitterrand's slogan—"Missiles are in the East, and demonstrations in the West"—because of its crudeness, fitted perfectly into Reagan's propaganda.

The alignment of the French president was not really surprising, since it had long roots. As he himself stresses, in the vital choice facing the French after the Second World War, he was not neutral; he chose the United States rather than the Soviet Union.[2] Thereafter, he was always one of those leftist advocates of European integration who took the American domination of the Western bloc more or less for granted. And he had criticized the deployment of the SS-20s before he became president. What was astonishing was the speed and the open zeal with which he pleaded Reagan's case, not minding being seen doing so. In an interview in *Stern,* the German weekly, given within weeks after taking office, Mitterrand proclaimed the need for American weapons "to redress the balance."[3] He repeated his plea time and again, his campaign reaching a climax on January 20, 1983, when he addressed the Bundestag in Bonn. The solemn occasion was the twentieth anniversary of the Franco-German treaty signed by Charles de Gaulle and Konrad Adenauer. Mitterrand told his audience that the allies must stand united—that is, accept the Euromissiles—if they wished the SS-20s to be withdrawn. He also warned that "whoever would gamble on the 'de-linking' between the European continent and the American continent would, in our view, upset the balance of forces and, hence, the maintenance of peace." Thus on the eve of Germany's general election, France's Socialist president was publicly backing the champion of the Right, Chancellor Helmut Kohl, and criticizing his Social Democratic opponents. In October 1983, he went out of his way in Belgium to provoke antinuclear protesters with the taunt that "pacifism is in the West, and Euromissiles in the East."

Mitterrand, who would not have the missiles on French soil, was not judging the matter from a distance. He was taking sides on the spot, in countries where the weapons were to be deployed. His bias was unmistakable. His own position, he claimed, had been reached after consulting the "best experts," who had reported that the military balance was tilted in the Soviet Union's favor in conventional forces, in tactical weapons, and, at least until 1985, in strategic armaments. To say that the experts were far from unanimous and that Mitterrand had decided in advance who were "the best and the brightest" is an understatement. Besides, Mitterrand then chose the arguments to emphasize. The deployment of Euromissiles was introducing a qualitative modification. With Pershing-2 missiles stationed in the Federal Republic, Soviet targets could be hit within five or six minutes, whereas it would take the Russians at least twenty minutes to reach objectives in the United States. Remembering the Cuban missile crisis, one can imagine the outcry if the Soviet Union were to place medium-range missiles close to the frontiers of the United States. The French president was perfectly well aware of it. Writing in 1986, well after the critical moment, he did not hide it: "In modifying to its advantage the time factor, NATO, in order to reestablish the tactical balance, dealt a severe blow to the strategic balance."[4] In the heat of the battle, he preferred to neglect this aspect and put the emphasis on the quantitative change.

Western Europe as a whole had been within reach of Soviet missiles, strategic, intermediate, and tactical, well before the advent of the SS-20s. The replacement of the medium-range missiles by the new ones, while possibly not altering the nature of the problem, undeniably increased the power and precision of the Soviet weapons, and everybody was entitled to ask the reasons why. The trouble with Mitterrand is that he overemphasized one aspect of the question while minimizing others. The proof of his special pleading was contained in his oversimplified slogan. Had the French president argued that missiles are on both sides, yet genuine, mass antinuclear demonstrations are, for the time being, not tolerated in the Soviet bloc, he would have initiated a real debate, raising an issue that the antinuclear movement itself had no right to ignore. Instead, he chose to score a propaganda point and offer Reagan a convenient quotation.

It is more difficult to determine why Mitterrand endorsed the American case so swiftly and so thoroughly. Because of the biases in his arguments, real fear of the Soviet Union's domination of Europe cannot provide a full explanation. Some charitable critics say that the French acted in this way from fear of German reunification, which is their main obsession. (As François Mauriac, the novelist, put it wittily: "I love Germany so much that I want to see two of them.") In this version, they wanted American missiles to perpetuate the American presence in Europe and, with it, the division of Germany. One flaw in this reasoning is that this was not the only solution. Perturbed at one stage by Germany's power and its opposition to his policy, General de Gaulle had turned to the Kremlin rather than to the White House. A more pedestrian motive certainly did play a part. Aligning himself with Reagan in what appeared at the time as the main front, the Socialist president could include Communist ministers in his government, pay lip service to movements of national liberation, and still be seen in Washington as a man beyond the suspicion of betraying the "Atlantic" cause.

The essential reason may be simpler still. Mitterrand may

not have really wanted parity but his side to keep the upper hand, and he knew where he belonged. Paris may quarrel with Washington over monetary matters and quibble over Central America, but in the foreseeable future these will remain family squabbles. If this interpretation is true, it raises something more fundamental. On reflection, Mitterrand never seriously contemplated his country's future in terms of nonalignment, of a Socialist third force, on a European scale to begin with. Deep down, he did not quite believe in his own verbal fireworks of the 1970s. He was intrinsically unable to look beyond the capitalist horizon. The domestic conversion and the foreign alignment strengthen each other, springing from the same source.

There remains the question of how in a sophisticated country like France, Mitterrand could get away with such crude propaganda and, behind it, the bigger mystery that had puzzled outsiders: Why was France an island in Europe, somehow by-passed by the mighty tide of the antinuclear movement? While other countries were deeply split, torn apart by a passionate debate, France, at least on the surface, looked aloof and unperturbed. It has been argued that the pacifist movement was strongest in mainly Protestant countries, with their "low-church" traditions of conscientious objection and their moral rejection of war as evil, whereas in France the antiwar movement was traditionally more political and revolutionary, with weapons supposed to be turned against officers or exploiters,[5] and radicalism was now at a low ebb. However interesting, the argument is insufficient, since the antinuclear movement was much stronger than in France in countries like Italy and Spain, which have a similar tradition to the French. Thus to seek a tentative answer to this major question, we shall have to look briefly at recent history—the Gaullist heritage and the ideological permutations of the Left—and at the distorting effect of the media.

Since General de Gaulle had taken France out of NATO's integrated command in 1966, it had no American bases or troops on its soil. There was no question of stationing Euromissiles there and no direct resentment against an obvious foreign finger on the nuclear trigger. Actually, some French citizens may even have foolishly assumed that should it come to a conflagration, nuclear weapons would distinguish and respect frontiers. This was one series of reasons why the Euromissile crisis provoked so little fuss in France. Another had to do with the recent conversion of the French Left to the nuclear creed. For a long time, the parties of the Left had been united in their criticism of General de Gaulle's nuclear strike force, his *bombinette,* dismissed as both useless and dangerous. The attacks, however, did not rest on a coherent or united policy of substitution—many Socialists, although obviously not the Communists, attacking Gaullism from a Euratlantic viewpoint—and when it approached the corridors of power, the Left accepted the military nuclear heritage, first as an inescapable fact, and then as an instrument that must be perfected and kept up to date. The Communists shifted first, in May 1977, seeing in the Gaullist strategy a potential weapon against American domination. The Socialists, whose change of mind had been prepared by men like Charles Hernu, of future Greenpeace fame, and Jean-Pierre Chevènement, followed suit eight months later. Even after this conversion, the defense strategy of the Left was liable to differing interpretations, but the advent to office of a united Left officially committed to atomic weapons was a serious handicap for the antinuclear movement.

To put it differently, the weakness of that movement was

a sign of the collapse in France of the unorthodox New Left. France did have after 1968 the embryo of a movement similar to those that were to grow in other countries, opposed to the civilian as well as the military use of the atom on moral and political grounds, seeing in nuclear developments a danger to humankind's survival and to the environment and, earlier because of its institutional implications, the threat of a police state.[6] This budding movement was gradually squeezed as the Left of the Common Program, obsessed with economic growth, traveled the road to electoral victory, and this is why the non-Communist contingent in the French peace movement had been so weak. The ambiguities of the Communist contingent are the second half of the same story.

The Communists were the only party of importance in France hostile to Euromissiles. But, as we just saw, they were in favor of French nuclear arms. They also belonged to a government keen on Pershing or Cruise missiles, which led to odd ventriloquism, Georges Marchais publicly endorsing Mitterrand's speech in the Bundestag.[7] Besides, the Communists had their own predicament. They knew that they could not attack Western Euromissiles without including the SS-20 in their condemnation, and since they had no desire to be tough on the Russians, they limited their campaign to a general waffle about the horrors of war and the desirability of peace. Last but not least, the Communists were distrusted because of their more distant past and their recent stand on Afghanistan and Poland, and thus many people were reluctant to join a peace movement dominated by the Communist party. It is true that the Communists were able to attract many more people to peace gatherings—a mass demonstration in Paris organized by the Committee of One Hundred on June 20, 1982, and a rally at Reuilly in the suburbs a year later—than to other meetings sponsored by them. But globally, France, far from being a leader, was a laggard in the antinuclear movement that was marching across Western Europe in the early 1980s.

As far as can be gathered, the lag was not really due to the mood of ordinary French people. According to comparative opinion polls, notably one carried by Louis Harris in 1983, the French were among the Western Europeans most frightened by the prospect of war and feared a nuclear conflict as much as the British.[8] Since in Britain the debate on the subject had been passionate, its conspicuous absence in France must be attributed, in addition to reasons already mentioned, to a more ruthless and successful filtering by the media. The matter was considered serious enough by most papers to leave it in the hands of trusted, orthodox "experts," and the antinuclear spokesmen had few regular outlets beyond the marginal Communist press. The bias was particularly striking in radio and television. Thus *The Day After,* the American film on the aftermath of nuclear bombing, was shown on television in most European countries. French television, all public at the time and with three channels at its disposal, could find no room for it. Since the film was not banned, it was shown in a Paris movie theater. Public radio thereupon organized a program around it, with high-school students invited to ask questions about the risks of nuclear doom. To answer them, it picked just one expert, ex–air force general and former employee of Dassault aircraft company Pierre Gallois, the most articulate and passionate advocate of the French nuclear force. (It was like allowing Dr. Strangelove to answer queries on the desirability of the bomb.) When French television decided to devote eighty minutes to the atomic question, on January 4, 1984, the screen was monopolized by André Glucksmann, new philosopher turned worshiper of nuclear power, and other upholders of the official line. The critics, the Communist Pierre Juquin and the retired admiral Antoine Sanguinetti, were allowed just to make an appearance. *Le Monde,* unable to conceal the glaring unbalance, tried to justify it on the grounds that Glucksmann's views "sum up the consensus of French political forces, with the exception of the CP." The preacher of pluralism was thus echoing the argument once used in Moscow to justify the eloquent silence of the dissidents.

Within such a context and in such a political climate, with the Right requiring no convincing and the Left on this occasion compelled to follow its government, Mitterrand had no difficulty backing Reagan in the battle of the Euromissiles. This initial community did not last forever. The French government had its own interests, and when these were hurt, notably by the American "Star Wars" project, Paris did not mind opposing Washington. But this initial association on a crucial issue did affect Mitterrand's image in the outside world.

10

In the Footsteps of Predecessors

In foreign policy, as in domestic policy, François Mitterrand's reign began with gestures and illusions. At the first Paris air show he visited as president, military planes were discreetly veiled, presumably to convey symbolically the idea that arms sales were part of a conservative past, not a Socialist future. In October 1981, on his way to Cancún for an international economic conference, Mitterrand stopped in Mexico City, where, drawing on the common revolutionary past of France and Mexico, he addressed the Third World as the champion of social justice "without which there can be no political stability." Close to the American frontier, he proclaimed that "East–West antagonism cannot explain the struggle for emancipation of the 'damned of the earth.' . . . Zapata and his men had not waited for Lenin to seize power in Moscow to take up arms against the unbearable dictatorship of Porfirio Díaz. . . ." And he went on to his peroration:

> Greetings to the humiliated, to the emigrants, to the exiles on their own soil, who want to live and live free. . . .
> Greetings to the women and men who are gagged, persecuted, or tortured and who want to live and live free.
> Greetings to the sequestered, the disappeared, the murdered who only wanted to live and live free.

In the Footsteps of Predecessors

Greetings to the priests, to the jailed unionists, to the jobless who sell their blood to survive, to Indians harried in their forests, to workers without rights, peasants without land, resisters without weapons, who want to live and live free.

To all of you France says: courage, freedom will win. . . .[1]

Stirring stuff and pronounced within hearing of the downtrodden of Latin America. And yet the message got a weaker echo in the Third World than had similar speeches delivered by General de Gaulle from Peru to Phnom Penh. Not because Mitterrand did not offer an alternative policy or economic means to carry it out. Neither had the general. But de Gaulle's challenge to the American leadership, however ineffective, had been perceived as genuine. Mitterrand's act was interpreted as empty gesturing or, as the southern French put it nicely, as "verbal words" (*paroles verbales*).

The instinct of the Third World listeners was not unsound. The early signs did not herald a new policy. Covering up military planes did not prevent France from preserving its third place, behind the United States and the Soviet Union, in the international league of arms merchants. If the number of its shipments declined at one stage, it was because its best customers, the Middle Eastern oil producers, were having financial difficulties.[2] The venture into Latin America was short-lived. The French president, although expressing it politely, did not conceal his disagreement with the American position: Revolutions are the outcome of domestic injustice, not a Soviet export; if they go the Cuban way, it is largely as a result of American bullying and blindness. In December 1981, the French actually signed a contract to supply Nicaragua's Sandinista government with two coastal boats, two helicopters, and some fifteen trucks.

The shipment was symbolic rather than significant, and then the whole enterprise fizzled out. The French did not change their principles; they simply ceased to put them into practice. Washington looked at this whole affair in proper perspective. French lip service to liberation in Latin America, which did not really count, was more than compensated for by its support of Euromissiles, which did matter.

At international conferences concerned with underdevelopment or with financial problems, France's president or its other delegates always presented a three-pronged program to help the have-not nations: an increase in public aid (which France did raise from 0.36 percent of its gross national product in 1981 to 0.5 percent four years later); better credit terms for the developing countries, thanks to an increased international liquidity; and regular prices for their exports, thanks to stabilization schemes. The French did not carry much weight within conference halls and not much conviction outside their walls. To stir the "damned of the earth" into action, more was needed than a project of relatively moderate exploitation. In the first couple of years of their rule, the Socialists were restrained by their wish to fit into Washington's schemes and get its blessing. Then, after its inner conversion, France ceased even to pretend that it was offering a different model and that its experiment could serve as an example.

For all his journeys, Mitterand's field of potential influence was thus shrinking. Verbally, he did show signs of political courage. In Moscow, to which he finally traveled in June 1984, he dared to mention publicly the name of Andrei Sakharov while speaking at a dinner in the Kremlin. In Israel, which he had visited much earlier, in March 1982, addressing the deputies in the Knesset, he talked of "the dialogue that assumes that each side can go to the limits of its right, which, for Palestinians like for others, when the time comes, may mean a state."[3] (But he could afford to speak so, since his visit was, symbolically, putting an end to a chapter of cool relations between France and Israel inaugurated, in 1967, by General de Gaulle's decision to lay an embargo on arms for Israel.) In the many Arab capitals he visited, he pleaded for the right of Israel to live "within secure and recognized frontiers." All this may have sounded magnificent, but it somehow did not add up to a policy. A medium-size state, unable to impose its line, can have an independent influence on the course of events in an area only if its proposals, corresponding to deep interests, strike a chord among local people. In the Middle East, France had a presence rather than an impact.

In Lebanon, where its links were old and strong, France did send soldiers. It already had troops in the south in the so-called United Nations Force of Intervention in Lebanon (UNIFIL). After the Israeli invasion, it shipped another contingent there, which joined Americans and Italians in August 1982 and helped in the evacuation of some 4,000 Palestinians. Early September, this multinational force left Lebanon, its peace-keeping task apparently accomplished. Ten days after its departure, a bigger French contingent was back. In the meantime, world public opinion had been shocked by the massacre of Palestinian civilians in the camps of Sabra and Shatila. Lebanon, to whose daily bloodshed the world had become accustomed, hit the front-page news again a year later, on October 23, 1983, when the explosion that cost the lives of 241 American marines was followed within minutes by the blow-up in which 58 Frenchmen perished. Although French troops stayed on for a time and the French air force carried out, two months later, an unsuccessful reprisals raid, bombing two Shiite camps near Baalbek, the goal of the Paris government toward the end of Socialist rule was to get French hostages back and no longer to shape the destiny of Lebanon. In the Middle East as a whole, the role of the French was reduced essentially to that of an arms merchant, particularly as a supplier of Iraq in the Gulf War.

French Socialists who originally had hoped, if not to lead, at least to inspire the Third World were rapidly reduced to exercising a real influence outside Europe, very much like their predecessors, only in France's backwater, its former empire in Africa. True, they could claim that, unlike President Valéry Giscard d'Estaing with Emperor Jean-Bedel Bokassa, they had suffered no major scandal, although this was probably the only serious claim they could make. The Socialists did not inaugurate an era, did not carry out any major reform marking the development of formerly French black Africa and likely to affect the rest of the continent. President Mitterrand and his advisers had to spend much time and energy to decide how far they should go acting as a police force in the area. Like the previous governments, the Socialists got bogged down in Chad.

More than twice the size of France, Chad has about 4.3 million inhabitants, among the very poorest in the world. Its interest is mainly strategic, enhanced by the fact that Colonel Muammar Qaddafi's Libya is its northern neighbor. An artificial product of colonialism, Chad was torn by tribal struggles. To prop up the regime, General de Gaulle had had to send troops there in 1969, as did President Giscard d'Estaing in 1978. When Mitterrand took over in France, there were Libyan troops throughout Chad, and the man nominally in charge in N'djamena, the capital, was Goukouni Oueddei. A few months later, the Libyans having left, the capital was taken over by Goukouni's old-time rival

Hissène Habré, the man favored by the Americans and, rumor had it, by the CIA. The defeated Goukouni took refuge in the sparsely populated north, looking to Libya for help. In 1983, the battle resumed in the north. In Paris, the lobby of ex-leftists now seeing the hand of Moscow strangling freedom all over the globe managed to build up an atmosphere of frenzy in the media, if not in the country. Washington was also urging France to intervene. Mitterrand hesitated. To intervene was to appear as a neocolonialist gendarme. Not to do so was to antagonize quite a number of African leaders who were relying on French arms to keep them in power. The second argument prevailed, and on August 8 the French launched Operation Manta, involving 3,200 troops and the latest equipment. The purpose was to keep Goukouni and his Libyan allies above a specified line, along the fifteenth parallel, which they were not supposed to cross; after a French plane was shot down in the no man's land, French troops moved the line to the sixteenth parallel. The protected area covered all the economically relevant regions of the country; in the north lay the vast desert with only 150,000 people. All this was fine but costly, and when Qaddafi let it be known that he would welcome a deal, the French jumped on the occasion. Claude Cheysson, the minister of external affairs, arrived in Tripoli on September 15 to finalize and sign an agreement on mutual evacuation of French and Libyan troops, which he summed up succinctly: "They stay, we stay; they go, we go; they have come back, we have come back." Both sides were supposed to have left the territory of Chad by November 15, when Mitterrand met Qaddafi on Crete. In fact, the Libyans had either left behind or reintroduced some troops, and the Americans, with their superior spying devices in the area, were delighted to report this breach of the agreement. The

French opposition and all those in Paris who saw President Reagan as the savior of Western values raised accusations of blindness and betrayal. Then the fuss died down. The French were satisfied with keeping Goukouni and the Libyans beyond the sixteenth parallel at a cheaper price from a distance with a warning line and the threat of reprisals. Although French involvement in Chad was very far from finished, Paris could turn its attention to more serious things.

If the story so far has given the impression of perfect continuity between the Right and the Left in the conduct of foreign affairs, then it must be shaded. The transition was much more complex, as can be seen in an adventure mixing the domestic with the distant. New Caledonia is officially French territory, although this Pacific island, or rather islands, lies between Australia and the Fiji archipelago some 13,700 miles from Paris. The native Melanesians, known as Kanaks, accounted for about 47 percent of the population of 132,000. The inhabitants of French origin, called Caldoches, whose numbers had grown particularly since the French departure from Algeria, represented nearly 40 percent. The rest were immigrants, mainly from Asia and Oceania, brought in to work on the estates or in the nickel mines that for a long time had ensured the prosperity of the territory. It was a typical colonial heritage, the invaders having grabbed the best land and driven away the Kanaks. After the end of the Second World War, France decided to give a degree of autonomy to New Caledonia, but in the 1960s the Right concluded that the population balance allowed France to go back to direct rule and block any roads that might lead to independence. The Socialists thought otherwise. They started talks with everybody, including the Kanak National Liberation Front (FNLKS). Edgard Pisani, ex-minister of de Gaulle and former commissioner in Brussels, was sent to Noumea, the capital, to negotiate, to stage local elections confirming the power base of both sides, and to prepare the ground for a new statute, which Mitterrand was to describe as "independence in association with France." The president, who visited the island in January 1985, promised a naval base as a gift for this limited independence. Because of clashes and conflicts, the referendum could not be held while the Socialists were still in office. The government of Jacques Chirac had no time for the FNLKS and no sympathy for any form of independence. In that sense, there was a real difference between Left and Right. On the other side, they both agreed that France had a divine right to sites for nuclear tests in the Pacific and that it was perfecly entitled to build a naval base thousands of miles from home. On balance, the element of continuity is, probably, stronger than that of contrast. There can be no question of a sharp break.

This small digression out of the way, we may ask whether the limited influence of French Socialists overseas was not due to their weak base at home and to the absence of a serious European substitute. General de Gaulle had already found, to his sorrow, that France on its own had neither the size nor the power to allow him to deal on equal terms with the superpowers. He therefore made the bid for the leadership of the countries "bordering on the Rhine, the Alps, and the Pyrenees," only to find out that this coalition, and Germany in the first place, could not be used as an instrument in the trial of strength with the United States. Mitterrand was to make the same discovery when he, in turn, if one may so paraphrase George Canning, tried to call in the Old World to redress the balance of the New.

As this implies, the Socialist president's alignment with the United States was not permanent, nor did it rule out clashes while it lasted. As early as June 1982, Paris, followed by London, Bonn, and Rome, decided to defy Washington by refusing to apply the embargo on supplies for the Soviet gas pipeline. European newspapers at the time were full of editorials telling Washington that the high principles applied to American farmers were equally valid for engineers on the other side of the Atlantic Ocean. All these were skirmishes. The serious clash of interests came later over Washington's so-called Strategic Defense Initiative (SDI), a project that was launched in 1983 but that became a bone of contention between the two countries in the next couple of years. The idea of a foolproof shield protecting the United States, thanks to a combination of lasers, satellite stations, and ground-to-air missiles, may well have been, as the nickname "Star Wars" suggests, a flight of fancy belonging to science fiction. What was true of the Soviet Union, with its monstrous arsenal resembling the American, did not necessarily apply to France, with its comparatively modest number of missiles.[4] If the Americans were to proceed with the SDI, they were likely to break the 1972 treaty limiting the development of antiballistic missile systems and thus precipitate a dramatic acceleration in the arms race, leaving France no longer able to reach the "threshold of terror," on which its strategy of deterrence rests. Naturally, "Star Wars" was to offer scope and profits to a number of firms in the arms industry—this was, indeed, one of its primary purposes—and a similar operation was too big and too costly for a country like France. Faced with the prospect of parading with an obviously obsolete nuclear sword, France was, not surprisingly, the first allied country openly to oppose the American project.

In order to counter the enthusiasm of European firms dazzled by the vision of a fall-out from SDI in the form of future juicy contracts, Mitterrand in April 1985 invented the European Research Coordination Agency (Eureka), a loose structure that was open to all countries of Western Europe and designed to ensure progress in advanced technology (notably in powerful lasers, big computers, new materials, and advanced robots). Although, in principle, this civilian enterprise had nothing to do with the American military project, Charles Hernu, still minister of defense at the time, talked of "star peace," and Roland Dumas, the new minister of external affairs, inviting other countries to the founding meeting of Eureka, pointed out that Europe would thus be able to speak on equal terms with its great partner: "A Europe of subcontractors, a Europe working under license would not be Europe." The response was polite rather than eager. France's partners joined the new club, the Germans even promising some money, but their membership did not prevent them from welcoming the SDI—the British wholeheartedly, the Germans with some reluctance. Mitterrand, in turn, was discovering that, organized as it was, Europe was not a card that could be used in a game with Washington. It remains to be seen why his European policy had to result in such a fiasco and to compare his with General de Gaulle's performance.

11

Europe and the Nation-State

Whereas Charles de Gaulle, deeply nationalistic, had resigned himself to a European solution out of necessity, François Mitterrand was, to use a cliché of the period, a "convinced European." He belonged, as he was fond of saying, to the generation that was born during the First World War, fought in the Second, and was exhilarated by the idea of reconciliation. He served in the government of Robert Schuman and expressed admiration for Jean Monnet, two of the French fathers of European integration and two advocates of a European evolution within the Atlantic orbit. True, as he assumed the leadership of the Left, Mitterrand became an occasional critic of American imperialism, and he got converted to the Gaullist doctrine of national deterrence. Europe, however, remained close to his heart and figured high on his agenda when he was elected president. He made eloquent speeches on the subject, acted as host in Paris to his European partners, and frequently visited their capitals, particularly during the first half of 1984, when, in keeping with the rotating system, France—that is to say, himself—exercised the presidency of the ruling Council of the European Economic Community. Considering the time and energy devoted to the EEC, the surprising thing is that when one looks at the balance sheet of the community during the five years of Socialist rule in France, nowhere does it bear the stamp of a Socialist president in one of the main member states.

Not that Mitterrand was inactive. On the contrary, he carefully supervised all the transactions, and important French interests were often involved. For more than a couple of years, the agenda of the EEC was dominated by, or rather encumbered with, Margaret Thatcher's "just return," Britain's claim for compensation because its contribution to the EEC budget was out of proportion to its national wealth. The French, while not minding transitional payments, refused the principle of a just return or proportional contribution. The British were paying a lot because they were importing a great deal, particularly foodstuffs, from outside the EEC. To ensure them a permanent and important compensation would have broken the basic rule of "community preference"—that is, reward for buying within and penalty for purchasing outside. France managed to rally its eight partners and isolate Britain on this issue.

As a counterpart, the French had to accept a tightening of the EEC budget. To cut the deficit, the contributions to the common budget from national taxes were slightly raised (from 1 percent to 1.4 percent of the value-added tax by January 1, 1986, with more to come). The key lay in the limitation of an expenditure getting out of control and, since the bulk of the money was spent on farming, in the gradual elimination of the notorious mountains of butter and other agricultural surpluses. In 1984, for instance, the EEC was expected to produce 104 million long tons of milk and to consume 85 million tons. So in April of that year, prices and subsidies were fixed in such a way as to reduce output and, progressively, to narrow and then eliminate the gap. The French, with their powerful farm lobby, were very vigilant in this negotiation because they were involved in the talks over a third item that aroused passion within the EEC, the so-called monetary compensations. These taxes or bonuses were the direct result of a myth. The common agricultural policy, the only really joint policy in the community, had been introduced in 1962 on the abstract assumption that the national currencies of the member countries would not change in relation to one another. This was just a fiction. When the franc was sharply devalued in 1969, since food prices were calculated in a European currency (which was not yet called the European Currency Unit, or ECU), they should have risen sharply in France. To avoid inflationary pressures, it was decided to leave them artificially unchanged; but to prevent French food from being sold cheaply in, say, Germany, French food exporters had to pay an equivalent tax—a "negative compensation payment"—at the frontier. When later in 1969 the German mark was revalued, the reverse process had to be carried out in order to avoid a big fall in the prices paid to German farmers, and a bonus or subsidy—a "positive compensation payment"—had to be given to the German exporters. Designed to facilitate a transition, the mechanism turned out to be a provisional that lasts. It was in the interest of a country like France, whose currency tended to be devalued in relation to the mark or the Dutch gilder, to have the whole system dismantled, and the decision to do so in stages was taken in 1984. As long as the currencies within the community fluctuate differently, it may well be a Sisyphean task.

All the measures mentioned so far were really accounting ones, which could have been taken by ministers and experts, with heads of government only giving their blessing. Two more ambitious steps were taken during the five years of Socialist rule in France. The first, initiated before the period, was the extension of the EEC from ten members, Greece having joined in 1981, to twelve, with the official admission of Spain and Portugal on January 1, 1986. The French had to ensure a transition period to safeguard the interests of their southern fruit and vegetable growers and to prevent this further expansion from turning the EEC into a mere customs union, a vast free-trade area without common policies. The second project, drafted during this period although enacted later, was concerned with the institutions of the community and the establishment of a single market. The EEC was a child of prosperity, and by breaking trade barriers, it had contributed to the rapid growth of production during its first fifteen years. The economic crisis had brought about a stagnation in the development of the community as such, and efforts were now being made to resume its forward march. On the institutional level, it was being proposed to increase the number of issues decided by a simple majority[1] and thus reduce the cases in which unanimity is required. It was amusing that the French should be the inspirers of this move, since the extension of the veto power had been imposed on reluctant partners by General de Gaulle in 1966 through the so-called Luxembourg compromise. The scope of the veto was now being restricted primarily to speed up the removal of various trade barriers in order to achieve within seven years "a space without frontiers," a genuine common market for 320 million consumers. The so-called Single European Act, embodying these amendments to the Rome Treaty, which established the EEC, was drafted when the Socialists were still in office and was enacted, in November 1986, by the time Jacques Chirac had taken over. There still remained the gap separating the act from its implementation in 1992.

Europe's record during the five years of Socialist rule in

France is neither nil nor particularly spectacular. Whether Valéry Giscard d'Estaing, had he been reelected, would have done better or worse than Mitterrand is, in a sense, a question of subjective assessment of the skill and determination of the two men. What is more difficult to see is what Miterrand achieved that his predecessor was intrinsically unable to do; in other words, the extent to which a Socialist president did defend different social interests. His only apparently left-wing proposal was the project submitted by the French in October 1981 to set up a "European social space," in which the Ten would coordinate their policies to reduce unemployment, to improve labor relations, and to harmonize their systems of welfare. To paraphrase a French saying, the alleged mountain did not give birth to even a sizable mouse. It was not Mitterrand's fault, it may be objected, since the French had to reckon with other members within a community dominated by conservative governments. The answer is that the French did not present a very radical plan, did not fight for it, did not attempt to attract the people of other nations in order to exercise pressure on governments, and quite rapidly stopped pretending that they were preaching by example. The Europe of states and of big business had made infinitely more progress than the Europe of labor unions and of ordinary citizens. No wonder that by the time Mitterrand called on his European partners to stand up to America, it was hard to see in the name of what they were to respond positively to his appeal.

Have we not just said that big business had prospered within the European Economic Community? Capital, by its very nature, breaks barriers and is patriotic only when it is profitable. Industrial concentration within the Common Market proceeded first along national frontiers and everywhere reached the stage where giants were swallowing one another. Cooperation then developed along European lines when it was directly sponsored by the state, such as in satellite launches with Ariane or in information technology with the European Strategic Program of Research and Development in Information Technology (Esprit), a forerunner of Eureka, or was simply encouraged by subsidies and orders, such as in the aircraft industry with Airbus. Otherwise, transnational mergers were in some cases confined to the frontiers of the Common Market; in others, they were stretching across the ocean. When a big French, British, or German firm was ready for a deal across frontiers or was simply ripe to be swallowed, it preferred to choose as a partner another firm likely to offer it additional markets and, above all, one possessing the most advanced technology. In fields such as electronics, computers, and telecommunications, more often than not such a partner happened to be American or, with the passing of time, Japanese. There was good capitalist logic in this attitude, and it was not going to be altered by capitalist or, for that matter, Socialist exhortation.

In the early years of the Common Market, there were passionate divisions between the advocates of a loose confederation of nation-states, for Gaullists a "Europe of Fatherlands,"[2] and the champions of supranationality, of a European federation. As the economic crisis continued to cripple a community whose numbers had doubled, federalism, the prospect of a European state, receded. Federation, it was said, requires somebody to impose it, and the three potential contenders were either unable or unwilling to do the job.

The Federal Republic of Germany seemed the most likely candidate to do for the European customs unions what Prussia had done for the German *Zollverein,* and as the

European Monetary System looked more and more as a sort of Deutsche mark area, the odds were on politics following the pattern of economics. Things were not quite so simple. Although Germany was no longer a "political dwarf," it remained a divided country. As long as its regime accepted the permanent division of the world into two blocs and the necessity of an American nuclear umbrella for its own survival, West Germany was in no position to bid openly for the leadership of Western Europe.

France cherished the illusion for a brief spell during the reign of the general, the shadow of Charles the Tall concealing for a time the economic imbalance with its eastern neighbor. De Gaulle's successors had only the nuclear sword as their claim to leadership, and it was increasingly unwieldy as a political weapon. The idea occasionally put forward of combining the French and British nuclear forces into a kind of European command vanished on its own when Britain remained a nuclear power by courtesy of Washington. Mitterrand, for his part, while reviving the provisions for military consultations between the two countries contained in the Franco-German treaty, proclaimed quite plainly that the French deterrent could conceivably be used only to protect the "national sanctuary" or a "vital interest," as defined by the president. Nuclear umbrella for nuclear umbrella, as long as the German government considered that it needed one, the American one looked more serious.

The United States was the third potential challenger. In the 1950s and early 1960s, in the era of Konrad Adenauer, Alcide de Gasperi, and Robert Schuman and shortly after, American diplomacy did envisage a united Western Europe as a junior partner. The image fashionable at the time, the two pillars of the alliance, was a rather shaky one, since the partners were not destined to be really equal. In fact, Washington dropped the project altogether because of Europe's growing strength in economic competition. A closely knit Western Europe coordinating its investments and pooling its resources in research and development would have become a seriously dangerous rival. The Americans decided to rely on their military command and on Europe's lack of unity to perpetuate their domination. Forever? General de Gaulle's struggle against this hegemony, his anachronistic realism, gives us a good background to understand Mitterrand's performance.

With the unification of Germany and Italy, the nineteenth century had been for Europe the golden age of the nation-state. The twentieth century, at least its second half, was supposed to be the era, if not yet of a world order, at least of vast aggregates the size of the United States or the Soviet Union. This seemed to be in keeping with the progress of science and communications, the economies of scale, and the need of big markets for mass production and for heavy investment in research as well as equipment. De Gaulle's realism was to have grasped that however obsolete historically, the nation-state still offered plenty of scope for bold action. He took France out of NATO's integrated military command. He interrupted the process of European integration. He questioned the role of the United States as an international gendarme. He challenged its position, admittedly only with words, in its Latin American backyard as well as in Phnom Penh, next to the Vietnamese battlefield. He did it and got away with it, showing by contrast the cowardice of other European politicians.

It was a dazzling act and, in a certain way, no more than an act. Removed through the main door, the Americans were returning through the window in the disguise of multinational firms. De Gaulle the nationalist was unable to resist the internationalization of capital, and "the defense of French grandeur" was not a slogan likely to rally Europeans en masse to the rescue. Some Socialist critics at the time argued that only a left-wing government would have the capacity to fill the "grand design" with substance, to turn the ultimately empty gestures into a policy.[3] By nationalizing big firms, it was to thwart foreign takeover bids and reorganize the methods of production. Through democratic planning, it was to build up the country's defenses against foreign invasion, at least for a period. Meanwhile, such a Socialist regime, groping toward new ways of producing and consuming, toward different patterns of life would have a chance to attract the people of Europe, if not their current rulers. A Socialist government came and did nothing of the kind. Far from dwarfing General de Gaulle, Mitterrand retrospectively enhanced his stature. The Socialist president did nationalize some conglomerates and banks, only to treat them as capitalist enterprises. Far from developing the planning system and rendering it more democratic, he dismantled and discarded the few instruments available for guiding the economy. Thus he had little to offer to the people outside France and actually no intention of appealing to them above the heads of their governments. To put it plainly, he did not even try. The still unanswered question relevant to outsiders as well as the French is whether he would have stood a chance had he tried.

12

Socialism and National Frontiers

Can one imagine a medium-size state like France or Britain making a radical break with its international environment and beginning to build a socialist alternative? The operative word is *beginning*. The concept of full-fledged socialism in a country was an aberration in Stalin's time, over sixty years ago, in a vast nation stretching over Europe and Asia. In our interdependent world, it is absurd in theoretical as well as practical terms. Socialism is international by definition, and its vocation is worldwide.[1] The question is simply whether the process that the French, before 1981, used to call "the break with capitalism" can start within the confines of a nation-state not greater than Italy or Germany.

A positive answer is not as obvious as it would seem to be. After all, it is no longer possible to conceive of a radical transformation of, say, Massachusetts or Yorkshire or Piedmont alone. Their respective nations are sufficiently centralized that if the change did not spread at once to the United States or Britain or Italy as a whole, it would be crushed by economic and, if necessary, military means. The European Economic Community has not quite reached that stage. Power does not have to be seized yet in Brussels, except, naturally, by the Belgians. The frontiers of France or Britain still provide the space for the beginning of a social experiment, although not for very long. At the pace at which the economies are becoming interwoven, this proposition may no longer hold true at the beginning of the next millennium, which is just round the corner.

Not for very long either in another sense. The borders of the medium-size state are too cramped for the experiment to last. They offer merely a breathing space, a few years in which to spread or surrender. This scarcity of time combined with the compulsion to convert neighbors forces the transformation into a narrow road full of pitfalls and contradictions. Take as an example the need for sheltering the economy during the transition. The case against protectionism is not only that it slows down progress in the long run. It also may immediately produce a drastic cut in living standards, which would not render the survival of the new regime by democratic means any easier. And the international division of labor is not innocent. It dictates the pattern of production, the size of firms, the methods of organization of work, the very terrain the new regime must tackle if it is to introduce new forms of democracy and attack alienation at its roots. Besides, what is true of production and trade is also true of culture and commercialization, of films, television satellites, and other means of communication. A socialist government attempting to reshape society needs some shelter to get on with its task. At the same time, it must minimize external shocks bound to aggravate domestic upheavals.

Let there be no illusion. However legitimate, democratically chosen, and popular, a regime that does not respect the established rules of the game will be violently attacked, opposition at home relying on foreign support. But the newcomer would not be without defense mechanisms. Take France as an example. It can invoke special circumstances and the "safeguard clauses" of the Rome Treaty in order to introduce legally provisional measures of protection. And beyond the law lie the interests. If France is tied to its trading partners, its own exit from the Common Market would precipitate the latter into crisis. Allow for inner conflicts, varying attitudes toward the United States, new opportunities offered by the Soviet market as well as the potential of different relations with the Third World, and there is quite a lot of scope for bargaining and room for maneuver, provided that all this remains marginal and transitional. Ultimate safety lies in permanent expansion. Not so long ago, one would have written that the socialist experiment would spread, say, from France to Italy, from Italy to Spain and Portugal, only thereafter moving to Germany or Britain. Today, the order of precedence is not so easy to forecast. But the impossibility to stand still remains, and this imperative need to spread has its own consequences.

One is to rule out a nationalistic response. A few years ago in the British Labour party, a member's radicalism was measured by his or her degree of hostility toward the Common Market. The funny side was that left-wing opposition to a wider unit made sense only on the grounds that a radical program would clash with the liberal framework of the European Economic Community, while nobody suspected the future Labour government of harboring such radical intentions. More seriously, we do know where jingoism within the labor movement can lead—to the blowing up of the *Rainbow Warrior,* to the patriotic stampede at the time of the Falklands War, to open hostility toward immigrant workers and well beyond. Now internationalism has acquired a new incentive, the inevitability of future collaboration between these medium-size states. This should have meant that it was no longer wise for left-wing parties in any

Common Market country to take protective measures or oppose further integration in the name of nationalism. Such operations should have been presented or carried out in the name of the future united socialist states of Europe and backed by the promise to accomplish with the fellow workers the elimination of frontiers that was refused on employers' terms. In practice, capital has shown much more capacity for rising above frontiers than has labor. International collaboration among unions should have proceeded much more rapidly. Whenever the workers of a nation ask for higher wages, shorter hours, better working conditions, or more welfare, the refusal of or simply the resistance to their demands is justified, whether by the state or by private employers, in terms of foreign competition and for members of the EEC in terms of the direct competition of the nation's partners. Thus it is astonishing that the labor unions have not staged so far a massive Continental campaign, say, for the introduction of the thirty-five-hour week, to begin with, throughout the EEC, that they have not been fighting together for the highest common denominator in vacations, health services, unemployment benefits, or old-age pensions.

Naturally, should they do so, it would be objected that they would price themselves out of markets compared with countries outside the community. The socialist experiment, once it started spreading, would know no frontiers. Even the European Economic Community would be only a stage in its development, and, incidentally, the united socialist states of Western Europe, too, might require an outer tariff, a protective wall, until they could win the economic comparison with the United States. Only the calculation would not be made in the classical manner, by dividing per head of population a gross national product, including a hefty proportion wasted on weapons and a smaller dose accounted for by advertising spent to convince us to choose between two similar toothpastes produced by the same firm. A socialist Europe and a capitalist United States would have to be measured in terms of use as opposed to exchange value. The comparison would not be limited to average consumption per head without taking inequalities into account. It would involve leisure as well as labor; not just productivity, but also conditions of work; and, ultimately, the degree to which women and men have a mastery over their fate. Once the overall balance were tipped in favor of socialist Europe, the need for a protective curtain would change sides. Yet, well before, a Western Europe forging a different society would act like a magnet for the Soviet bloc and would inevitably establish radically altered relations with the people of the Third World in Latin America, Asia, and Africa. The United States could not remain unaffected.

Let us not get carried away by a distant possibility, a glimmer of hope. What the reasoning behind this scenario reveals is the link, from start to finish, between the domestic and the foreign. The socialist experiment in Western Europe can survive only by spreading; its salvation lies in the conquest of new territory, but it can advance only by example, by its force of attraction. The missing unifier of the European Economic Community may be the country and class offering a genuine social alternative. The historical task of a socialist government, whether in France, Britain, or Italy, is not to fit into the international division of labor by trying to make the best of it. It is, while opposing its own logic to that of the capitalist system, to weather the international storm long enough to be able to exercise its powers of attraction. The wider questions, which we shall briefly review in Chapter 15, are whether socialism can be brought up to date, reinvented in order to appear as clearly relevant to the solution of major contemporary problems, and whether it can shed its bureaucratic, dictatorial disguise in order to become attractive once again.

Meanwhile, in the last section of this chapter the reader may have gained the impression that the French president was somehow lost. He was not, but was lurking behind every line of the argument. To analyze what a protagonist might have done and has not is sometimes more revealing than to describe what he actually achieved. The connection between Mitterand's performance on three stages is now established. He was unsuccessful in his bid for a special position as spokesman for the Third World; he was then unable to harness Western Europe to oppose even a mild resistance to American projects, because he did not offer even a semblance of a radical alternative at home. The failure of the traveler and the diplomat was rooted in the failure of the domestic reformer.

Yet is it fair to speak of failure at home when many people, including probably many readers, see there his monumental achievement?

IV

Lessons from France

> The trumpet of a prophecy! O wind,
> If Winter comes, can Spring be far behind?
> Shelley, "Ode to the West Wind"

If beauty is in the eye of the beholder, the verdict is in the mind of the judge and depends on his or her own code of values. Accused of treating François Mitterrand harshly, the author is entitled to plead that the socialist standards applied to his performance had been, after all, Mitterrand's own. Yet it is true that Western politicians are not normally measured with such exacting yardsticks. Compared with his immediate French predecessors or with the "Iron Lady" across the Channel, Mitterrand does not come off badly. Seen from a totally different angle, the socialist principles discarded, he rises in stature. He is the man who has precipitated the fall of the French Communist party. He is, or at least looks like, the president who stripped the French Left of its revolutionary heritage and dragged it into the "normalcy" of Western politics. Should this achievement prove lasting, which remains to be seen, he certainly deserves a place, if not in the socialist pantheon, at least in some capitalist Hall of Fame.

13

The Case For and Against Mitterrand

Projections over time and space offer plenty of room for cheating. By choosing a convenient base year or ignoring a context, it is easy to distort the whole picture. Comparing François Mitterrand's France with its neighbors and predecessors, we shall try not to read too much into the figures in order not to lie with statistics. The first thing to reemphasize is that comparisons with the United Staes make no sense, if only because France was in no position to finance its external deficit by printing money. When in 1982 French exports paid for only 80 percent of the country's imports, its whole economic policy had to be revised at once. The year after, American exports paid for only 78 percent of the import bill, and by 1985 the ratio was down to 62 percent.

During the five years of Socialist government, the French gross domestic product rose by barely 6 percent. The annual average of 1.2 percent was around the mean for the EEC, on a par with that of Germany, and below the British rate. In industrial output, France, which showed no progress over the period, lagged behind both of its European competitors. But it did better, or rather less badly, in unemployment. The jobless accounted for 6.3 percent of the French labor force in 1980 and for 10.1 percent in 1985, having risen by an annual average of 12 percent. In the Federal Republic, their share rose from 3 to 8.6 percent, or by an average of 37 percent. In Britain, it climbed from 6.4 to 13 percent, or by an average of 20.2 percent. Looking at these bare figures, one might get the wrong impression that France was keeping up with the Germans and lagging behind the British. The reality was very different. Britain was flattered by the figures because its cycle had started earlier; 1980, the base year, had been one of serious slump. While the difference between the French and the German rates of inflation was smaller at the end than at the beginning of the five years, France was running a big trade deficit in manufactured goods with the Federal Republic. During the period, Germany was the only country of western Europe really to consolidate its industrial base. By the time the Right got back into office in France, the country was threatened with the new version of the "British sickness," the dangerous contrast between the prosperity of financial capital and the dramatic shrinking of industry. Leaving these prospects apart, the main feature of the figures for the five years was their similarity. The French pushing ahead first and dragging their feet afterward, it all came roughly to the same thing, which, depending on your political taste, may be interpreted as showing the capacity of the French Socialists to manage or their inability to invent.

Comparisons over time in France are even more controversial, since they lead straight into electoral polemics. The Socialists insisted on the contrast between the beginning and the end of the period in suitably selected sectors. The Right stressed the deficits accumulated over the period and chose fields that favored its own propaganda. It put the emphasis on the budget deficit, which, rising fivefold in five years, more than doubled the public debt from 418 billion to 1,068 billion francs, and on the continued climb of taxes and levies—state, local authorities, and welfare contributions all lumped together—from 42.5 to 45.5 percent of the gross domestic product. It added that for the first time in years, real wages actually declined in 1983 and 1984 and that the Left, which had promised to find outlets for the unemployed, in fact reduced by 600,000 the number of available jobs. The Socialists would rather talk about the rate of inflation, in the region of 14 percent when they came into office and approaching 2.6 percent when they left it. They retorted that the domestic state deficit, including local authorities and welfare, at less than 3 percent of the gross national product, was well below the average for Western Europe and the United States, that foreign trade was nearly balanced, and that the net foreign debt was roughly the equivalent of six weeks of French exports. The Right, they added, better be quiet on the rising tax burden or on the growth of unemployment, since both had increased during the presidency of Valéry Giscard d'Estaing twice as fast as during the Socialist government.[1]

Such statistical back-and-forth activity can go on for long, with each side simply scoring points. This is not the heart of the matter. If after the Socialist interlude Raymond Barre became the most popular conservative leader, judging by opinion polls, this was not because of his economic record as Giscard's prime minister. At that time, he had reached rock bottom. He subsequently recovered because the Left had restored *his* credibility. It had sworn that his remedy was useless and that it had a radically different, real cure, and it finally dished out a similar medicine.

Similar does not mean the same. The Socialists, as they reached office, expressed their predilections. They raised the wages of those with the lowest pay as well as old-age pensions and family allowances. At the same time, they introduced a wealth tax and raised the highest rate for income tax. The Left also sponsored two important social reforms: the fifth week of vacation with pay and the possibility of retiring with a full pension at age sixty. Although the Right attacked this legislation and voted against it, once back in office it did not dare to brave unpopularity by reversing it.

There were less specifically economic measures distinguishing the Left from its predecessors, and for the final judgment on Mitterrand's reign to be fair, we must now briefly recall some of the liberal and progressive steps taken by the Socialists in their years in office. One can thus mention the abolition of the death penalty and of all exceptional jurisdiction, the liberal handling of justice by Robert Badinter, its minister, and the spur given to scientific research by Jean-Pierre Chevènement and his successors at the ministry bearing that name. Jack Lang, responsible for culture, doubled its share in the budget and used the funds cleverly to encourage writers, painters, musicians, and filmmakers. The establishment of the *ministère des droits de la femme,* with Yvette Roudy in charge, while not bringing any revolutionary improvements in the condition of French women, showed at least a better awareness of their predicament.[2] Some measures were also taken to render the fate of immigrants more bearable: Notably, a foreigner could no longer be expelled by the whim of the administration—that is, the police—a decision of the court henceforth being required for the purpose.[3] The government also encouraged a movement such as SOS-Racisme, whose button stating "Keep Off My Pal" flourished all over the country and which managed to stage a Woodstock in the heart of Paris, bringing some

300,000 young people to an antiracist concert in the place de la Concorde on the night of June 15, 1985.

The picture of the Left is more rosy in retrospect than it actually was, and the contrast is sharper when it is set against the image of its successors rather than that of its forerunners. The conservatives who returned to office in 1986 were incomparably more aggressive, reactionary, and sure of themselves than they had been in 1980, and this because of what the Socialists had done, and not done, in the meantime. The Left had thus contributed to the renewed contrast and to the retrospective glow, a point that may be better grasped by looking once again at the case of nationalization.

When the Left was bidding for office, nationalization was presented as an instrument for greater efficiency and for social transformation. By the end of the Socialists' term, the only possible justification for the takeover was the utter failure of private enterprise to do the job. Meanwhile, so much had been done to praise the private and damn the public, bestow a legitimacy on profit or the stock exchange that in its platform for the 1986 election, the Right dared to propose denationalization lock, stock, and barrel, with the followers of Jacques Chirac claiming the Gaullist heritage and promising to turn into private hands the big banks and the Renault works nationalized by the general. And the Left during that campaign, you may recall, merely retorted that this program of "privatization" was based on ideological rather than economic grounds. By the time the new government began to fulfill its pledge, the leftish whimper became even more plaintive, the only thing wrong with the dismantlement of the public sector being the estimation of its value: The newcomers were selling the family jewels at bargain-basement prices. It is on this fantastic revaluation of values, this staggering change of mood in so short a time that the case for or against Mitterrand really rests.

If the future of Europe is unavoidably American, capitalism the only possible horizon, and the search for an alternative an illusion, if one not only accepts this but also, bowing to the inevitable, welcomes it, then one can praise a man who, having grasped the historical necessity, found a shortcut to take France, the rebel and outsider, into the resigned European pack. Whether or not they are spelled out, these are the necessary preconditions to demand a monument to Mitterrand the normalizer. Even so, one should not indulge in the cult of personality: It was not just his own work. We tried to show in Chapter 2 that this passage to resignation was not in Western Europe a sudden jump. It was a long process made up of bitter disappointments, the collapse of the Soviet model, the economic success story of Western capitalism, and the disarray of the Left when the capitalist fiesta came to an end. In the more specific French context, it involved the upheaval of 1968 and the need for years to contain potentially explosive forces. By the time Mitterrand became president, quite a lot had been achieved. But the experience would not have been complete without the Left in office publicly demonstrating its own impotence to alter the established order.

Thus it did not start with Mitterrand, nor did he do it on his own. Mitterrand needed the Communists badly to help him in his exercise. Their role had been much more important than his in 1968; now it was to be subordinate and still significant. Mitterrand required both their blessing and their antics, the former to show that the whole Left was participating in the experiment, the latter for the Communists to be acting simultaneously as a bugbear. The fact that the

Communists, as long as they remained in the government, described its record as superior to that of the popular front and, on the morrow of their departure, branded the government as an obedient tool of the bourgeoisie did not enhance the credibility of the Communist party, and this helped to discredit the idea of an alternative. To recognize other contributions is not to deny that at the heart of the transformation stood the Socialists and their leader, who had plenty on their plate.

In the first place, they had to impose a series of concessions on the French labor movement. Probably the easiest was the freeze and even a slight cut in real wages. Coming after the initial lavishness, it might have been easier still to accept, had it been accompanied by an inspiring project of social change. Instead, it was coupled with a scheme for "restructuring" French industry, a process that the Right had not quite dared to launch and that, affecting the main concentrations of labor, was hitting the unions where they were strongest. Last but not least, the workers were being deprived of some of their guarantees, such as the sliding scale for wages, and toward the end of Socialist rule, they were urged to sacrifice many of their conquests on the altar of flexibility. The conservatives probably would have carried the offensive over the same terrain, but they would have had to fight a bloody battle. The Left managed to do it while preserving "social peace," keeping the hours lost annually as a result of strikes at exceptionally low levels.[4] The price for it all was paid by the further weakening of the already not very powerful unions. With only about one fifth of the labor force belonging to them, the unions were stronger in militancy than in numbers. During the years of left-wing rule, they were to suffer on both counts, the two more radical unions being the main victims of association with the government. The General Confederation of Labor (CGT) followed a cycle parallel to that of the Communist party: It began by containing discontent on the shop floor, and, then, after the Communists had left the government, it tried to send the workers to battle by whistling orders, with an understandable lack of success. The French Democratic Labor Confederation (CFDT) succeeded in identifying workers' democracy with concessions to the employers, and thus its own setback coincided with at least a temporary discrediting of new, original ideas associated in the French movement with *autogestion*.

The retreat on the ideological front was even more impressive. The late chairman of Schlumberger, Jean Riboud, would have had reason to be proud of his friend, who had managed, at least on the level of the media, to reconcile the Left and private enterprise. Profit was being presented as virtuous in itself rather than as a measure of capitalist success, and the stock exchange as a new temple of French society rather than a symbol of casino capitalism. The shares on the Paris Bourse, incidentally, after an initial fall in 1981, rose by 145 percent in the next four years, faster than on most other stock exchanges and about twice the pace of shares on the New York Stock Exchange. Although the ideological sea change, as we saw, had begun earlier, the Left's years in office accelerated the trend beyond recognition. The contrast mentioned earlier between Jacques Attali's view of socialism *before* and that of Michel Rocard *after* could be multiplied ad infinitum with quotations from Mitterrand himself, from Left and Right, from government and opposition, from politicians and leader writers. The shift in the center of political gravity had been so rapid that it was often hard to believe that the contrasting statements had been made by the same man or the same party. Still, the Socialists are probably the most striking example. In 1980, "social democrat" was among French Socialists a title of insult referring to somebody guilty of "class collaboration." Five years later, a congress of the German Social Democratic party was a "hotbed of Reds" by French standards. To say that the French Socialists had mellowed is an understatement. The concepts of class and capitalism, even the very word *socialism,* had disappeared from their vocabulary. The epitome of radicalism at a Socialist meeting, and the party's only remaining link with the Left, was the assertion that social justice and solidarity are necessary in order to carry out the indispensable economic reforms. Such subversive stuff would not send a shiver down the spines of delegates to an American Democratic party convention.

A bitter lesson of realism, a cure for utopian illusions, and the shattering of mythologies—these are some of the arguments used to explain and justify France's fall into line, the shedding of its original belief in the possibility of radical change. As "destroyer of dreams," Rocard is small fry. Pride of place must be given to whom it belongs, to the French president—distant, dignified, saying the opposite with the assurance of a man who has never moved an inch. International business, however, was not to be fooled by such an appearance of continuity. When Mitterrand was elected, he was viewed with suspicion. Five years later, when the defeated Socialists were leaving the government, the lonely president was being hailed unanimously by *The Economist, The Financial Times,* and *The Wall Street Journal* as a wise statesman, having done yeoman service for the Left, for France, and for the Western world. To say that these are not exactly the journals usually bestowing proletarian honors is irrelevant. The days of seeking Socialist laurels were well gone for Mitterrand. The doubt was over the kind of reward he would receive from the capitalist Establishment. Did he deserve a monument for a historical achievement or merely a chocolate medal for services rendered, useful but ephemeral?

14

Crisis and Polarization, or For Whom the Bell Tolls

François Mitterrand brought his party to the middle of the road, into the realm of compromise and consensus, his survival as president with a hostile parliament confirming this evolution. In the United States, the coexistence between the president and Congress rests on the agreement between the two sides, the Democrats and the Republicans, on certain fundamental points about existing society. Conflict and compromise are also written into the Constitution, with checks and balances ensuring the functioning of the system. In France, on the contrary, the president can bring the confrontation to a head at any moment by asking the verdict of the people. The fact that Mitterrand, despite that opportunity, stayed in office shows how much water he had poured into his red wine. While a bipartisan approach to foreign and defense policy is quite frequent in the Western world, much more was involved in France. A Socialist president limited his opposition to a few polite words of disapproval, while the government over whose sessions he presided announced its intention, for instance, to privatize everything. Although perfectly entitled to stay, the president was not, in the French context, compelled to do so. His choice extended the limits of the tolerable beyond belief. A late starter, the French Left was entering the world of consensus with a vengeance.

Zeal does not always lead to permanence. The desire of the Socialists to stick to the new line is obvious, and doubts about their capacity to do so are linked only with the new economic era in which they must perform. The consensus, the acceptance by the main political protagonists of the established society and of its basic rules, was in other countries a product of history, consolidated in Western Europe after the Second World War by thirty years of unprecedented growth. France joined in much later. Nobody still seriously pretends that the troubles affecting the world economy since the 1970s can be blamed on the wicked Arabs and their high price of oil. There is broad agreement about the gravity of the current structural crisis affecting capitalism, with big differences as to its outcome. Will the crisis slowly deepen as increases in productivity result in higher unemployment? Will it be dramatically accelerated by a financial collapse? Or, with nobody pushing it off the stage, will capital continue its reign by discovering new engines of expansion? Fortunately, for the sake of our argument, we do not need a clear answer to this fascinating speculation. Choosing the assumption most favorable for it, the system will require quite a few years of transition. This "restructuring," which has already begun, should enable it to destroy values by closing down plants and to improve the rate of profit by cutting labor costs and lifting restrictions imposed on capital in the years of prosperity.

When a social system that has functioned apparently smoothly runs into obvious trouble, it becomes vulnerable, and this is when it must convey the impression that there is no better way out. While the Soviet Union, painted as the incarnation of evil and a warning, was ideal for this purpose, Mitterrand's France was very precious, too. What better than a leftist government could, by its blundering, demonstrate the vanity of the very search for something else. The glee with which its failure and conversion were greeted in the Western media and conveyed to the public was only too obvious. But it was easier to score such propaganda points than to deal with another consequence of the economic crisis: By increasing social tensions, it tends to polarize politics.

The "radical center," the golden mean, or, to borrow Valéry Giscard d'Estaing's contribution, "France wants to be governed in the middle" ("La France veut être gouvernée au centre")—for several years, politicians and pundits had been preaching the virtues of moderation, and a message drummed so methodically by the media was bound to have some effect on public opinion. Yet when you look at it more closely, they had a strange conception of compromise, favoring, if one may say so, a one-sided polarization. The pendulum was allowed to swing fully to the right; then in reverse, it was supposed to stop somewhere just left of center. Ronald Reagan and Margaret Thatcher were not historical accidents. Their personalities undoubtedly a factor, their election was essentially a reflection of the economic crisis and of the decision by key sections of the Establishment that the cost of the postwar consensus was by then prohibitive, that it had to be dropped altogether or shifted. As a result, their economic and social policies went well beyond the limits of the respectable conservatism of the postwar period. Logically, the movement to the right was to be accompanied or followed by a similar swing to the left, and this danger had to be avoided at all costs. The preaching was destined to only one side. Varying from country to country, depending on local circumstances, an attempt was being made throughout the Western world to narrow political choice from the right to just beyond the center: from Reagan to Hart in the United States; from Thatcher to the Alliance and, since that was impossible in Britain, to a Labour party chastened by the Social Democratic splinter; and from Chirac and Barre to sobered Socialists, preferably allied with liberal conservatives, in France. All this was fine on paper and could even last for a while, part of the electorate opting temporarily for a middle course after the excesses of the Right. It could even last for quite a time in the absence of a clear alternative. Yet fundamentally, the new construction was built on sand.

After all, the success of reformism in Western Europe following the Second World War, the relative moderation of the Left, and its acceptance of the social compromise were not due to the charisma of social-democratic leaders; actually, they were not charismatic. The "discreet charm of the bourgeoisie" rested on much more solid foundations: on steady and substantial increases in living standards; on sliding scales guaranteeing wages against inflation; on growing insurance against ill health, old age, and the apparently vanishing scourge of unemployment; on more room being provided for workers' children in an expanding educational system—in other words, on a whole complex edifice that was now being threatened by stagnating production and a volume of unemployment reaching proportions unthinkable since the war. French Socialists, following their British and German comrades, had to consolidate their new positions at a time when nothing seemed to favor or facilitate that consolidation.

Doubtless, the ideological posture is far from negligible. After all, it was essentially this heritage—the political class consciousness, the belief in the possibility of changing life radically through political action—that distinguished for quite a time the French and the Italian Left from their British and German counterparts. The existence of big Communist parties in the two Latin countries was partly the effect and partly the cause of this phenomenon. Yet it was not all heritage, culture, and ideology. Consensus had flourished, favored by special conditions, and these conditions were disappearing. The economic crisis was throwing a new light on the contradiction between our technical capacity to build a new world and our social inability to make it a place worth living in. Improvements in productivity, instead of spelling greater freedom, were translated into higher unemployment. While the extension of the robots was supposed to put an end to heavy, monotonous, or broken-up labor, alienating work in fact gained ground, spreading from the factory into the office. With the line of the jobless lengthening and the pressure to cut the social services rising, discussions about the way in which society should be run ceased to be theoretical or abstract. Debates, too, were inevitably polarized. The Establishment could no longer preserve its supremacy by preaching compromise. It had to revive slogans forgotten since the Second World War about the laws of the jungle and the survival of the fittest. It had to split, whenever possible, the employed from the unemployed, the natives from the foreigners; to oppose income differentiation to the principle of workers' solidarity; and to weaken the latter, wherever it could, by anti-union legislation. Even the welfare state, a few years earlier proudly presented as an instrument of social justice in an almost egalitarian society, came under attack as being too costly and harming economic development. Here and there, voices were raised in Europe to praise the American model of two-tier social services, with basic welfare for the really poor and private insurance for those who can afford it.

Faced with this continuing trend, Europe's left-wing parties, however social democratic, and its labor unions, however moderate, felt threatened in their very existence. The other side changing the nature of the game, at stake was their very survival as bodies defending, in their own way, the immediate collective interests of their members. No wonder that the stolid and steady Labour party seemed to be shifting really to the left at one stage, when Tony Benn just missed his chance, and although Neil Kinnock moved it back to respectability, the inner struggle is very far from finished. No wonder either that the German Social Democrats should tackle issues too radical for their French comrades. Incidentally, the rise of the Greens in Germany is a permanent reminder for the SPD that without a global vision, it has no chance of capturing the sympathy and backing of the young generation. While the Left in Europe's Latin countries, France included, is discovering the virtues of consensus politics, its older European practitioners are becoming aware of the limits of the system. Will the roles be reversed? Probably not. The odds are that, having joined a creed deprived of its attractions and a system losing its tempting assets, the converts will rapidly lose their zeal. Already, there are signs that the conversion is more superficial than the picture given by the media suggests. All these are reasons to believe that Mitterrand may deserve a modest medal rather than a lasting statue. But although the situation will vary from country to country, much will depend on the international development of the economic crisis and on the room for maneuver that it will leave for the capitalist Establishment. Bets on the future are, by their very nature, speculative.

But France, after five years of Socialist rule, teaches us a more immediate and tangible lesson about the scope available for reshaping our societies, although not necessarily the lesson usually drawn from the events. France is an excellent example because, with its Revolutionary memories and its radical traditions, with a united Left returning to office after twenty-three years of absence, the desire for change was unquestioned. We saw Mitterrand the liberal progressive converted to socialism, and we were entitled to some doubts about his rhetoric. When he wrote that the initial task of the Left will be "to break the chains of social inequality,"[1] when he talked of "destroying capitalism and its masters"[2] or proclaimed "that big business [*le grand capital*], master of levers of economic and political command, remains the enemy number one, with which there can be no possible compromise,"[3] we tended to take such pronouncements with a grain of salt. But it was not all hypocrisy and camouflage. The eagerness of the Socialists and of their leader, not to change society altogether, but to make changes within it, to carry out important reforms, is undoubted. Had they reached office fifteen or twenty years earlier, they might have gotten away with it and preserved some principles. As it was, they were ill-equipped to deal with their historical task. The Common Program offered no solution for an open society torn by the economic crisis. Since they were not pushed forward by a vast social movement, as soon as they met serious resistance, they surrendered. Contrary to common wisdom, the Mitterrand experiment tells us directly nothing about the possibility, or impossibility, of building socialism—that is, about the search for a radically different society, implying a break with its predecessor. It simply had not been tried. The French events shatter the hopes or illusions that the transformation can be carried out gradually, without any break, within existing institutions, by purely parliamentary means, without the active participation of the people in their factories and their offices, without the unconcealed vision of another world indispensable to produce such a mobilization. If something had to be sung as the French Left was being electorally buried in March 1986, it was not a funeral oration for socialism, but a requiem for social democracy.

15

Socialism to Be "Reinvented"

Whatever the pundits may have been preaching, the French experience does not prove that socialism will not and cannot work. It does not prove the opposite either, and very far from it. Just looking at what François Mitterrand and his ministers should have done and did not, one perceives the immensity of the task facing a government and a movement groping genuinely, although without undue illusions, toward a really different society. *Immense* is not synonymous with *impossible*. Indeed, it is rather puzzling that Marxism, treated with a distant respect a few years earlier, was being totally dismissed just at the time when its critique of capitalism, emphasizing the contradiction between productive forces and social relations as well as the inevitability of economic crises, seemed more relevant than it had since the Second World War. In Western Europe, for socialism as an idea the moment of lowest attraction coincided with that of highest opportunity.

On reflection, it is not so puzzling. In Eastern Europe, the bloody collectivization, the ruthless industrialization drive, Stalin's purges, the extension of the empire—all the vagaries of "primitive accumulation" were being presented as the essence of socialism and paradise on earth. In Western Europe, the reforms carried out by social democracy, relatively important in countries like Sweden, nowhere altered fundamentally the fate of humanity and the individual's control over society. Between the "really existing socialism" and what it stood for in the eastern half of Europe and its real inexistence in the western half, socialism had been reduced to the stature of a ghost. Like love in Arthur Rimbaud's famous sentence, socialism had to be "reinvented"; it had to be morally resurrected and brought up to date. I will not have the arrogance to pretend that I could do so if I had the space. Drawing on the French experience, I will simply try to list the issues that a socialist movement must tackle if once again it is to attract adherents and to appear capable of answering the questions of our time.

The first series of problems is connected with the state and democracy. To reveal the emptiness of current attacks on an abstract state, root of all evil, in the name of an equally abstract civil society, to show that the freedom now being praised is often the famous "freedom of the fox in the chicken coop," is all very fine and very insufficient. The apprehension of the mighty Moloch in Paris or in London is not exclusively, or even mainly, the result of the Soviet Union's unwithered state seventy years after the Revolution. Big Brother is resented closer to home as a bad boss, a snooping supervisor, and a distant, bureaucratic distributor of welfare. The socialist movement must, in this context, restate clearly its position from scratch. It is no admirer of the state as such, which it proposes ultimately to abolish. Nevertheless, it knows that the state will not vanish by some magic as long as the differences not just in income, but also in property, power, and privilege, on which rests the state and which it is designed to perpetuate, remain. While the "withering away" of the state is thus a distant goal, a socialist government must tackle it from the very start. A nationalized industry cannot be described as fulfilling its function because, for example, it is more competitive in foreign markets. Nationalization, or better still socialization, does not make sense unless it brings about changes in the organization of labor leading to a greater mastery of the working people over their jobs and their factories. Services, too, will not properly be called social until they are run not only for, but also under the effective control of their beneficiaries.

Democracy at all levels and the state are not linked accidentally. People's fate will not be altered without their increased mastery over their own working lives, which will involve the size of plants as well as property relations. Computers, data banks, and the like may well help to decentralize decisions, but they will not eliminate the need for some form of central planning. (Those who rightly point out a certain revival of small and medium-size firms in the past few years often forget that these have also been years of financial concentration, with a record number of takeover bids.) In any case, a central body will be required to coordinate supplies, to organize research and distribute its findings, and to act as a collective agent in relations with the outside world. For a long period of transition, a central-planning authority will be needed to uproot the inequalities among plants and among regions. To introduce democracy into the work place, a revolutionary innovation in itself, will not be enough. It will be vital to invent and improve new forms of political representation—with rotation, possibility of recall, a permanent flow of news and decisions in both directions—in order to transcend the present passive method of one vote every few years. It is thus, by supplying a social content and increasing active participation, that socialism can substantiate its superiority over the current system of marketplace democracy.

The second series of problems is connected with the labor movement and the role of "historical agency" that the working class plays in the socialist conception of the world. Here again, the easiest thing to do is to deal with what passes for conventional wisdom. For those who argue that the labor movement has no longer the muscle to play a leading part, it is enough to recall the events of Paris in May 1968 and of Poland in August 1980 where, by throwing down their tools and paralyzing the country, the workers turned ordinary plays into major political dramas. Those who say that the working class may well have been perceived as the main historical actor in Marx's time, but is no longer large enough for that function, will be simply asked to do their sums again. Numerically, and even proportionately, the industrial proletariat is bigger today than it was in the mid-nineteenth century. Meanwhile, the huge peasantry has shrunk beyond recognition, while the number of white-collar workers has grown and keeps on growing.

Scoring debating points is not a substitute for analysis. The deep modifications in the shape and inner make-up of the working class are an important fact. The feeling of alienation at work, once the prerogative of the industrial laborer, has spread to the modern office worker. With the very process of production being affected by the intensive technological application of science, the line between "productive" and "unproductive" labor has become more difficult to draw and can no longer be made to coincide with manual and nonmanual. The consequences of this metamorphosis, for trade unions in particular and the Left in general, were mentioned earlier, and here we only want to get on with the question whether the working class so transformed can and should still be seen as a potential instrument for the reshaping of society. In the socialist context, this assertion simply meant that, in addition to their social weight and political muscle, the workers as a class had a particular stake in the abolition of class society. Workers today, to paraphrase the cliché, have much more than chains to lose. They have cars and television sets and often their own houses. But they do not own the means of production. They have no social or cultural capital to sell, only their labor for hire. Thus they have an interest in getting rid of all privileges and can still be conceived collectively as the driving force in the struggle for the establishment of a classless society. To deprive socialism of this vision, which some will dismiss as messianic or utopian, of a joint struggle for an egalitarian, although in no way uniform, society in which exploitation will be ultimately uprooted and even the division of labor will change nature, would be to reduce socialism to an eclectic electoral cartel, content with speculating on various forms of discontent.

The next range of problems writes itself. Since politics, like nature, abhors a vacuum, the list of social movements that have sprung up or developed in the last quarter of a century reads like an indictment of socialism, an enunciation of its shortcomings and failures. At least two of them, because of their significance, must be noted: women's liberation and the ecological movement. Socialists, it may be pleaded, were among the first to draw attention to the exploitation of women, even to their double exploitation. In practical terms, they did not do a fraction of what they should have done, and tackled very timidly the discrimination at work and almost not at all the expression of male chauvinism at home and in society at large. Like the other oppressed, the blacks in the United States for instance, women had to act separately first in order to be able to act together, and despite some progress, it would be presumptuous to pretend that feminism has really permeated socialist culture.

Ecology in the widest sense of the term, going beyond the insertion of people into the environment, reveals the other big gap in socialist practice. Preoccupied with productivity and growth as such, large sections of the European Left talked and acted as though the means did not matter, as though once goods were plentiful, everything else would look after itself. Some of them failed to see in nuclear weapons what was always considered to be the ultimate alternative to socialism—barbarity pushed to its extreme form, nuclear doom. Wider sections still showed in the debate over the civilian uses of nuclear energy a curious insensitivity to the kind of economy that it implied and to the social consequences that the security precautions might involve. They revealed a lack of interest in the desirable size of the work place, in the ability of the associate producers to gain control over it, altogether in what is loosely called self-management, as though a truly socialist economy could be run just by orders from above. The emergence of the Greens in Germany is an obvious price paid for such socialist sins and a potential portent of things to come. If most feminists and ecologists cannot feel at ease within its ranks, Europe's socialist movement will not carry the day and, indeed, will not deserve to win.

There is still another way to grasp the void created by socialist unfulfillment, its degeneration in the East and abortions in the West. The success of Aleksandr Solzhenitsyn, the popularity of Pope John Paul II, and, although this bracketing will shock, the rise of Khomeini and the spread of fundamentalism in the Islamic world are all, in their own way, a measure of this historic failure. When reason fails to provide solutions, humankind, understandably, although dangerously, seeks them in irrationality. It echoes the Baudelairean phrase "anywhere out of this world . . ."

Socialism, by definition, cannot be treated in national isolation, and in Chapter 12 we tried to deal with its European dimension. A medium-size nation-state, we argued, offers a launching pad, at best a provisional platform. Socialists are thus condemned to internationalism here and now. The labor unions, in particular, will either extend their action to Western Europe as a whole or be progressively driven back to their infancy, to the function of factory unions coping with the interests of one craft in one plant. We left aside, however, the much bigger issue of the relationship between the socialists in the advanced countries—capitalist and post-capitalist, East and West—and the exploited, downtrodden, often starving millions in the Third World. Does their road to socialism lead, even in an accelerated course, through all the stages of our development, including its period of "primitive accumulation"? Can they find shortcuts? Should they seek different roads and therefore different solutions, which are bound to affect our own conception of socialism, still conceived, whether in Europe or the United States, from the point of view of the have-nots of the have nations? Finally, can one envisage a "socialist island," even if stretched through the whole of Western Europe, facing a sea of poverty—nay, surviving thanks to the existence of that poverty? To the last question, the negative answer is as simple as the solutions are complex. We have been able to only mention these problems here, adding that the way in which they are tackled will be vital, strategically as well as morally, for the future of the socialist movement.

There is also the original issue of socialism as a comprehensive and coherent alternative to capitalism. In the late 1960s, the ghost of revolution was haunting an apparently prosperous Europe and France in particular. Twenty years on, in a Western Europe saddled with 19 million unemployed, the established order reigns supreme. Its self-assurance, bordering on arrogance, is due to the absence of a serious opposition, to the resignation of the discontented, to the general mood, confirmed by the Mitterrand experience, that nothing will, that nothing can be done. An alternative, in this context, should not be seen as a detailed description of a different society and a precise list of all the steps to be taken to reach it. Europe is littered with splendid-looking programs and broken promises, so people will not be easily fooled by another catalog. But since the bitter Russian experience has taught them that there is no automatic, inexorable road from revolution to socialist democracy, the mass of the people will not be moved into action once again without a global project, without a clear vision of where they are heading, and without democratic safeguards for the road.

Why a project at all, since distant goals can lose a handsome number of immediate precious votes? Thus, for example, the egalitarian aspect of socialism may antagonize potential supporters who have relatively high incomes and, more generally, those who are on the right side of the threatened hierarchical divide. The answer here depends on fundamentals, on one's very conception of socialism. If socialism is not a present from above for passive and faithful voters, if it is a conquest of powers by the associate producers who gain mastery over their jobs, their society, and their fate step by step, each advance conditioned by the development of their political awareness, then cheating oneself into power is self-defeating.

Historical hiccups are not at stake here. There is no question of repeating in another place and another time the heroic storming of the Winter Palace by the Bolsheviks, a sudden seizure of power. It would be more appropriate to speak of a long march within and without the institutions, the ultimate seizure of power through the conquest of powers at all levels, the search for social and cultural hegemony. To stick together in such an expedition, an alliance based on common long-term interests and a common vision, is required, not a loose coalition of often conflicting discontents. To resurrect a credible alternative, the Left will have to raise its sights above the purely electoral horizon.

Not so long ago, it was fashionable in left-wing circles to talk of the need to get into the engine room of the capitalist economy and to lay a hand on the commanding levers. The metaphor was both deceptive and instructive. It was not much use and could even be dangerous to maneuver the levers of command differently unless one was ready to overhaul the engines drastically. Mitterrand and his comrades demonstrated that one could do better still—enter the engine room in the moment of crisis as a rescue team and, the takeover completed, continue roughly the same course. They showed the socialists of the Western world how to win an electoral battle and lose the political war.

16

The Negative Lesson

How to take office without seizing power. The French lesson is especially instructive because, unlike a Harold Wilson in Britain or a Helmut Schmidt in Germany, François Mitterrand was not openly committed to the preservation of the system basically as it stood. The French Socialists had thus written a sort of *What's Not to Be Done* for their Western European comrades who really intend to change the social and political shape of their continent.

The main lesson is that the fate of such an experience is decided well before the election and depends largely on the Left's capacity to mobilize and on the way it sets about it. Since a socialist government determined to keep its promise is bound to meet stubborn resistance throughout, it cannot ask its supporters for just a ballot paper. It must be brought to power by a vast social movement, which is not quite the same as being swept into office by a tide. As the government will need backing for sustained, long-term action, the movement must be politically conscious, aware of its interests and objectives. This is why the Left should not get into office under false pretenses or as a cartel of the discontented, why it should not play hide-and-seek with its program. Other social groups must be attracted by the project and the sweep of the labor movement, not by winks, innuendos, and purely electoral pledges. The difficulty we had in distinguishing whether the French Socialists before 1981 were for a "break with capitalism" or for its management was due to such ambiguity of language, which varied according to the kind of meeting and audience addressed.

It was not quite all over by the time of the election. Mitterrand, with his feel for politics, was right in asserting the primacy of the first few months, although wrong in interpreting this to mean that one should take advantage of this early euphoria to pass a few laws and spend the rest of the time digesting the reforms. The open proclamation of such a strategy is an advance invitation to a retreat. The first weeks and months are crucial because this is when the balance of forces is established and the momentum set. For the Socialists, this was the moment not so much to write articles with the heady title "To Govern Differently" as to do so in practice, to invent links with their supporters at all levels. Only a permanent dialogue with its rank and file as well as perfect clarity about the measures initiated and their long-term consequences would enable a socialist government to propose, should the situation so require, a compromise, a pause, even a temporary retreat. Or, more accurately, it would enable a vast movement publicly to debate how far to go and what risks to take. To object that such open politics gives the game away is to miss the point about a left-wing government. Conspiracy does not fool its enemy; it demobilizes its supporters. Let there be no illusions. Even a genuine and inventive socialist government backed by a movement gaining in political consciousness as it moves along would not have it its own way. Considering the domestic and foreign problems that socialists would have to face when beginning their experiment in a medium-size state, it would look more like permanent brinkmanship.

The alternative, as the French Socialists under Mitterrand so clearly illustrated, is surrender with or without abdication.

The lesson is important for Europe's socialists because everywhere they seem to be approaching a major confrontation, as Right and Left in succession have proved unable to cope with the economic crisis or, more precisely, to solve it to the satisfaction of the great majority of the people as opposed to the few who are making handsome profits. The confrontation is likely to affect the whole of Western Europe. Until now, in countries such as Spain, Portugal, and Greece, recently emerged from fascism, the Left had also a liberalizing function, for customs as well as institutions, which gained it a certain indulgence for its economic shortcomings. This is probably coming to an end. Everywhere, the Left will soon be judged essentially on its ability to control a country's economic development, symbolized by its level of unemployment.

Europe's political map is increasingly difficult to decipher as old stereotypes disappear. The Left in Italy, although less spectacularly than in France, is also losing its traditional belief in the possibility of radical change. At the same time, the faith in the virtues of gradualism is being eroded in the bastions of social democracy. The two warring factions of Europe's Old Left are in a serious predicament simultaneously. The Communist half, originally inspired by the October Revolution and then sterilized by Stalin and his heirs, is a provisionally spent force groping in all sorts of directions. Social democracy is without any project whatsoever, now that the gospel of capitalism's perpetual growth has been discarded. To complete the picture, the New Left, which thirty years ago had made a promising beginning, failed to grow anywhere to a politically significant size. Thus the paradox is that just when objectively everything seemed to be ripe for its revival, the Left found itself bereft of ideas and thrown on the defensive.

What it had to face were not the mild electoral skirmishes of yesterday. The Establishment apparently decided that the new period demanded a frontal offensive, and Margaret Thatcher's new Tories were merely zealous forerunners. One task of the Right was to recover a lot of territory ceded to the unions in the years of postwar collaboration. Another, exploiting the growing fragmentation of the labor market, was to divide the working people, to move from a society in which class solidarity was an accepted heritage to another in which only charity, preferably means-tested, would be tolerated. Even labor unions that had never dreamed of questioning, let alone attacking, the established order felt their very existence was endangered. And things remaining the same, the welfare state was next in line after the unions.

The confrontation was dangerous in yet another sense. Until now, although unemployment itself is demoralizing and degrading, the living conditions of the victims have not come down to the prewar level. As a much lower rate of growth and high unemployment have become lasting features of the Western economy, the present level of unemployment benefits, pensions, and health services is very far from assured. If an international financial crash were to come on top of it, hell would break loose. Even the relatively milder current crisis suggests what can happen when social tensions build up. The rise of Jean-Marie Le Pen and of xenophobia in France is a warning that, given the circumstances, our societies can run amok once again, and this time they have the nuclear means of collective suicide.

Although its responsibility is tremendous, can the Left come up on time with solutions? Curiously, all the relevant questions were raised in anticipation in the 1960s, although admittedly in a vague, utopian, and sometimes metaphysical way. Attacking the so-called consumer society, the protesters of that decade put the problem of growth at the heart of the debate. Growth for whom, for whose sake or profit, and for what purpose? Commodity production is not necessarily the same thing as the satisfaction of human needs. If the latter is the aim, it raises the issues of not only what to produce, but also how to produce—the questions connected with the size and organization of the production unit, the control of the producers over their work, and, beyond, the abolition of the frontier between manual and mental work, between labor and leisure. All these questions not only were unanswered, but were swept aside by the economic troubles. Who cares about job enrichment or about the associate producers' mastery over their working life when the main preoccupation, once again, is to get a job, any job?

The heady 1960s were followed by the pragmatic, corporatist, selfish 1970s. Factories were closed, costs cut, and profits raised, and after years of such "modernization" the system does not work any better. The face of Europe is actually more disfigured than ever by unemployment, and statistical cosmetics cannot conceal it. The questions raised in the 1960s, before the open economic crisis, are now much more topical than they were at the time. The Left, which stupidly allowed itself to be bullied off the subject, must rapidly invent concrete and clear answers. Once again, it must think boldly of different patterns of production and consumption and of a society in which "disposable time and not labor time" will be the measure of wealth[1] and in which progress of human creativity will, therefore, not be simultaneously a curse. If it does not do it swiftly, the clash between our inventive genius and our social as well as political backwardness may blow this planet to pieces.

Yet is there still any sense to talk of such a breakthrough or to think at all in terms of a historical future? The most fashionable word today seems to be the prefix *post-*. We are living in a "postindustrial" society and admiring "postmodern art." The trendy commentators bombard us with futuristic images of nuclear-triggered X-ray lasers, of one-world television beamed from satellites, of robots doing our work and computers our thinking. However, if you dare to ask why it is that a world changing so fantastically in so many respects must somehow be tied forever to the same forms of property and exploitation, you are dismissed as a dinosaur. On reflection, the philosophy behind this futuristic mumbo jumbo is rather old-fashioned. Like all ruling classes, the present-day one admits the existence of history up to its own triumph, although not beyond. Post-everything means capitalist forever. There was history, but time must now have a stop. Europe may still move up to the American model; the United States is, by some strange malediction, condemned to the same social state forever. Actually, since they require evil Russia as a permanent prop, our specialists in propaganda projected American capitalism and Soviet neo-Stalinism as frozen twins tied to each other for eternity.

The magic did work for a time. As the Soviet Union seemingly slumbered during Leonid Brezhnev's later years, while in Washington, London, and Paris history was apparently advancing backward, people's imagination was affected by a feeling that there might be no future. Then the spell began to break. In Paris, in the winter of 1986, two events showed that the real picture was not quite the one conveyed by the media and, the more important side of the same coin, that people were able to resist better than was assumed the systematic brainwashing to which they had been subjected.

Young French people were supposed to be self-centered

and determined to get on. The age of commitment, as we saw, was allegedly over and that of careerism had begun, Bernard Tapie, the pushy tycoon, replacing Jean-Paul Sartre as an inspirer. Everyone for himself, and winner take all; the American Dream at its very worst was described as their ideal. Yet when the government, taking this for granted, decided to dismantle the egalitarian side of French higher education by allowing individual universities to be more or less selective, to have varying and not uniform fees throughout the country, and to have their own rather than national diplomas, half a million high-school and university students took to the streets in Paris. In the biggest youth demonstration ever, they chanted against discrimination, against schools for rich and poor, against what was presented as an American future. These youngsters, whose political apprenticeship had been served in the antiracist campaign with the button saying "Hands Off My Pal," were now joyfully proclaiming in surrealist terms that they were not "carpets"— *pas la génération moquette*—not a Tapie generation.² It was not all fun. It took a lot of tear gas, one dead, and many wounded for the Chirac government to yield. The young people were only beginning their political education. But this egalitarian revolt took France by surprise. The protesters were so different from the generally accepted picture of conservative social climbers that the enraged editor of the magazine section of *Le Figaro* described them as infected with "mental AIDS." This gentle comment was a measure of the shock. And in that very same December, France was in for another. The forgotten working class suddenly reminded the country of its existence. Outflanking the unions and setting up local and national coordinations of strikers, the railway workers paralyzed the country's transport system, and although the strike lasted over Christmas

The Negative Lesson 293

and the New Year, it was not really unpopular. France was not quite the country it seemed to be.

Come to think of it, neither was the Soviet bloc. It had been giving signs of motion much earlier. In that glorious summer of 1980, the workers of Gdansk, entering dramatically the political stage, showed that the labor movement could alter the course of things in Eastern Europe too and shape events from below. Six years later, in Russia, came the long-awaited signs that the dismantlement of the Stalinist heritage was being resumed, with Mikhail Gorbachev starting from above a process bound to affect all the sections of Soviet society. Shaken from the top and from the bottom, Stalin's old empire, preserved for so long by bureaucratic interests and conditioned reflexes, is probably entering a period of profound upheaval.

Let us not read into these events more than they merit. A march by politically awakening students is not even a replay; it is not May in December. One strike does not announce the revival of a labor movement. The great Soviet saga will, at best, have its ups and downs, its advances and retreats. All these happenings of greatly varying importance, and many more, have one thing in common. They give a lie to the prophets of the end of history. Time need not have a stop. The earth is not standing still. *E pur si muove.*

Our final verdict on Mitterrand and the Socialist experiment in office is inevitably subjective and depends on our respective views on this very issue. If you consider that capitalism marks the end of history, that Manhattan 1990 is the inevitable model for Europe, then you will hail Mitterrand as a wise statesman. You may treat him as a maverick, as a born-again defender of the established order, but you must be impressed by his performance. To have taken over the French Left, still slightly dominated by the Communists and vaguely dreaming of another world to replace bourgeois society, and, overtaking Europe's Social Democratic parties, to have approached the American Democratic party as an example, this march, even uncompleted, is quite an achievement. Within your terms of reference, it turns Mitterrand into the man who brought France politically into the modern age and the French Left to its senses.

But if, like myself, you are not convinced that time must have a stop and history end with capitalism, if you believe that, should it escape nuclear doom, humankind will soon resume its monumental, unfinished, and uncertain struggle for mastery over its own fate, then you must lump Mitterrand with Wilson and Schmidt—a more interesting case because of France's special circumstances, but belonging to the same species—as politicians who, whatever their ambitions and claims, have at best slowed down, not shaped, the course of events, the groping search for a socialist future.

His seventy-first birthday just over, Mitterrand, basking temporarily in the glory of a constitutional monarch enjoying more pomp than power, cannot yet be dismissed as a political figure. As these lines are written, he is actually faced with a difficult choice. He can, his seven-year term over, retire to his country house in Latché to watch trees grow and write memoirs in his elegant prose. He is very keen on setting himself on record. He will probably try to shade the differences among the three stages of his political life, to play down the attacks on capitalism and its law of the jungle in the second period, as he has minimized his proclamations of French permanence in Algeria during the first. He is likely to be more successful in conveying an impression of continuity than in convincing his readers that he had achieved his main ambition of being a shaper rather than a plaything of history, a real actor rather than a performer.

The Negative Lesson 295

He can also choose to soldier on. He is still perfectly alert, a champion of political calculation, a master of electoral moves. Should opinion polls suggest that he stands a real chance, he may seek another mandate. He could thus resume his task, Sisyphean in my view despite its apparent success, of consolidating in France the capitalist consensus. Whichever he chooses, the political epitaph for France's first directly elected Socialist president is already written: People do not make history by proxy, at least not socialist history.

Notes

Introduction: Is the Future American?

1. Unless one includes in this category the Poujadists, about whom more on pages 154–55.
2. The expression, originally coined in *Les Temps modernes*, the monthly review directed by Jean-Paul Sartre, is an allusion to his play *La Putain respectueuse* (*The Respectful Whore*).
3. For my own, different, view on the subject, see Daniel Singer, *Prelude to Revolution: France in May 1968* (New York: Hill & Wang, 1970).

1. Changing France in an Expanding Europe

1. The figures are taken from European Community Commission, *Perspectives de la politique agricole commune: Le Livre vert de la Commission* (Brussels, 1985), statistical annex, table 3. See also *Agriculture des pays de l'OCDE: problèmes et défis des années 1980* (Paris: Organisation for Economic Co-operation and Development, OECD, 1984), table 21, p. 118. Naturally, these are orders of magnitude, one acre differing from another depending on the region, the land, and the kind of farming. For another classification and an excellent summary of the present situation of French farming, see Nicole Eizner, *Les Paradoxes de l'agriculture française: La Fin de la France paysanne* (Paris: L'Harmattan, 1985). For background, see Georges Duby and Armand Wallon, *Histoire de la France rurale*. Vol. 4: *De 1914 à nos jours* (Paris: Le Seuil, 1976).

2. In 1980, 75 percent of the farmers in the countries that belonged to the European Economic Community were over forty-five years old (European Community Commission, *Perspectives de la politique,* p. 12). In France, 40 percent of heads of farm households are expected to retire between 1983 and 1993. Just over 33 percent of all French farmers are working full time on the land (Eizner, *Paradoxes de l'agriculture,* p. 16).
3. See pages 113–14.
4. Luc Boltansky, *Les Cadres: La Formation d'un groupe social* (Paris: Editions de Minuit, 1982). Chapter 2 discusses the switch from the idea of *classes moyennes,* based on inherited property and family, to the American concept of the middle classes, based on comfortable living standards.
5. Before the Second World War, the word *employé* implied a higher status, which it no longer does.
6. For international comparisons, see *The Integration of Women in the Economy* (Paris: OECD, 1985).
7. For these and the following statistics concerning women, see *Economie et statistique,* nos. 171–72 (Paris: Institut National de la Statistique et des Etudes Economiques [INSEE], November–December 1984), pp. 13–20, 37–47; *Données sociales, 1984* (Paris: INSEE, 1984), pp. 28–33; and *Recensement général de la population, 1982* (Paris: INSEE, 1984), pp. 55, 93–96.
8. Only in Great Britain, where they come mostly from the nations of the Commonwealth, do the West Indians, Pakistanis, and Indians have the right to vote in parliamentary elections.
9. In the population census of 1931, foreigners accounted for 6.6 percent of the total population and 7.4 percent of the labor force. The corresponding figures were 6.5 percent and 7.3 percent in 1975 and 6.8 percent and 6.6 percent in 1982. Statistique générale de la France, *Résultats statistiques du recensement général de la population 1931* (Paris: Imprimerie nationale, 1935), Tome I, 3e partie, p. 94; Tome I, 5e partie, p. 104; and INSEE, *Recensement général de la population de 1982: les étrangers* (Paris: La Documentation Française, 1984), pp. 15 and 23.
10. Centre d'Etude des Revenus et des Coûts [CERC], Alain Foulon, ed., *Les Revenus des ménages, 1960–1984,* documents

du CERC, no. 80 (Paris: La Documentation Française, 1986), particularly table on p. 104.

11. Imports rose from over 11 percent of gross domestic product in 1955 to almost 13 percent in 1960, around 20 percent in the mid-1970s, and close to 25 percent in the mid-1980s. OECD Economic Outlook, *Historical Statistics, 1960–1985* (Paris: OECD, 1987), pp. 67 and 68; for the earlier period see *Le Mouvement économique en France, 1949–1979* (Paris: INSEE, 1981), pp. 93 and 95.
12. For an interesting study that suggests this idea in passing, see Jacques Marseille, *Empire colonial et capitalisme français* (Paris: Albin Michel, 1984).

2. The Left Bewitched and Bewildered

1. See, for instance, Fernando Claudin, *Santiago Carrillo: Crónica de un secretario general* (Barcelona: Edition Planeta, 1983).
2. In October 1984, *The Economist,* in big advertisements, proudly recalled that six years earlier, it had told its readers everything about Conservative preparation for the miners' strike. Indeed, in the issue dated 27 May 1978 it had leaked the so-called Ridley report, which, restrospectively, can be described as the Tory battle plan, outlining most of the measures taken to prepare for the strike.

3. Seeds of Defeat in Victory

1. This power is given to the president in Article 16 of the constitution of the Fifth Republic.
2. This was actually the second election by universal suffrage. The first was the election of Louis Napoleon, the future Napoleon III, on December 10, 1848. But then only men voted.
3. Many of the "leftist" leaders of the revolt of May 1968, whether Maoist or Trotskyite, were former members of the Communist Students Union (UEC), whose opposition to the party Establishment started during the Algerian war.
4. In the first ballot of the 1969 presidential election, held on June 1, the Socialist candidate, Gaston Defferre, obtained only 5.07 percent of the votes cast, compared with 21.52 percent for

the Communist candidate, Jacques Duclos. Michel Rocard, then candidate of the Unified Socialist party, got 3.66 percent; the Trotskyite Alain Krivine, 1.06 percent.
5. Among its members who were to play a role during François Mitterrand's presidency were Charles Hernu, Louis Mermaz, Roland Dumas, Pierre Joxe, Georges Fillioud (the future minister of communications), and Guy Penne (in charge of African affairs at the Elysée).
6. One distinction should be drawn. Whereas most old Stalinists had been true believers, the new philosophers were obviously time servers.
7. Richard Crossman, ed., *The God that Failed* (New York: Harper, 1949). The contributors included Louis Fisher, André Gide (presented by Enid Starkie), Arthur Koestler, Ignazio Silone, Stephen Spender, and Richard Wright.
8. Guy Hocquenghem, *Lettre ouverte à ceux qui sont passés du col Mao au Rotary* (Paris: Albin Michel, 1986), p. 51.
9. Georges Marchais addressed the Central Committee in a secret session on June 29, 1972, two days after having signed the Common Program. François Mitterrand spoke in Vienna on June 28, 1972.
10. François Mitterrand, *The Wheat and the Chaff*, trans. Richard Woodward, Concilia Hayter, and Helen Lane (New York: Seaver Books/Lattès, 1982), p. 179.
11. For a flattering description of the Second Left, see H. Hamon and P. Rotman, *La Deuxième Gauche* (Paris: Editions Ramsay, 1982).

4. One Character in Search of a Role

1. François Mitterrand's parents resented colonial expeditions because "they distracted the French from the revenge on the Rhine." This quotation, and all those attributed to Mitterrand, are taken from *Ma part de vérité* (Paris: Fayard, 1969). He was later to admit that he had a rather "Barrèsian" image of France. Maurice Barrès (1862–1923), novelist and writer, became one of the main spokesmen for French nationalism.
2. Quoted in "Franco par Mitterrand," *Le Nouvel Observateur*, 6 October 1975, p. 34.

3. Jean-Noël de Lipkowski, a former minister, remembers having seen Mitterrand in prayer, kneeling on the stone floor of the cathedral of Cologne in 1956 (Catherine Nay, *The Black and the Red: François Mitterrand—The Story of an Ambition*, trans. Alan Sheridan [San Diego, Calif.: Harcourt Brace Jovanovich, 1987], p. 29). Discreet on the subject, Mitterrand now gives the impression of an agnostic with spiritual leanings.
4. He practiced law for a time only after 1958, the early period of the Fifth Republic.
5. This definition of the Radical party, the story has it, was once given by a Socialist to Edouard Herriot, the famous Radical leader and mayor of Lyons, who allegedly replied that the Socialist party reminded him of a restaurant that was advertised in his home town as a "working-class restaurant—*cuisine bourgeoise*," a play on words because in French, it also means "home cooking."
6. Mitterrand, *Ma part de vérité*, pp. 27–36.
7. The irony is that the man accused of being his "Red" accomplice was none other than Félix Houphouët-Boigny, one of today's most conservative leaders in Africa.
8. The following quotations are from Mitterrand, *Ma part de vérité*, pp. 38–42.
9. Ibid., p. 42.
10. Quoted in Didier Buffin and Paul Guilbert, "Mitterrand se penche sur son passé," *Le Quotidien de Paris*, October 26, 1977, pp. 2–3.
11. Quoted in Nay, *The Black and the Red*, p. 205.
12. The case takes its name from a mysterious attempt to kill with a bazooka General Raoul Salan, then the commander in chief in Algeria. Salan escaped, but one of the officers on his staff was killed. Michel Debré, a passionate advocate of *Algérie française*, was mentioned among the alleged inspirers of the plot. For a subsequent version of the bazooka case by Mitterrand, who was minister of justice in 1957, see his testimony at Salan's trial in 1962 (François Mitterrand, *Politique* [Paris: Fayard, 1977], pp. 403–6). Salan was tried and sentenced to life imprisonment for his participation in the "putsch of the generals," staged in Algeria in April 1961, and for his subsequent leadership of the OAS.

13. Gaston Defferre was launched in *L'Express,* which was started in 1953 as a radical weekly by Jean-Jacques Servan-Schreiber and a group of leftist journalists. It backed Pierre Mendès-France and opposed the war in Algeria. At the end of that war, Servan-Schreiber decided that the moment had come for a magazine less committed to the Left, for a sort of French *Time.* Part of the staff then left the paper and contributed to the relaunching of another magazine, *Le Nouvel Observateur. L'Express,* having moved to the center, sponsored Monsieur X and Defferre's candidacy. Since then, bought and resold by Sir James Goldsmith, it finds itself very much on the conservative side of the French political spectrum.
14. Archives of the Socialist party, quoted in Mitterrand, *Politique,* pp. 531–42.
15. Mitterrand, *The Wheat and the Chaff,* p. 128.
16. Ibid., p. 194.

5. The Fall from Grace

1. Lang was helped by Serge Moatti, the film director.
2. The ministries were filled, respectively, by Marcel Rigout, Jack Ralite, and Anicet Le Pors.
3. Thierry Pfister, *A Matignon au temps de l'union de la gauche* (Paris: Hachette, 1985), pp. 98–99.
4. Quoted in Jean Marie Colombani, *Portrait du président* (Paris: Gallimard, 1985), pp. 209–10.
5. P. Alexandre and Jacques Delors, *En sortir ou pas* (Paris: Grasset, 1985), p. 71.
6. The European Monetary System (EMS) was set up in March 1979, after the collapse of the "snake in the tunnel," to preserve "an area of stability in Europe." In a world of floating currencies, it is an attempt to keep the currencies of the EEC moving more or less together vis-à-vis the outside world. The money of the members is defined in the European Currency Unit (ECU), a basket of EEC currencies weighted according to their country's economic importance. The rate of fluctuation of participating currencies among them is limited to a band of plus and minus 2.5 percent. When the bottom or the ceiling is

reached, the central banks are bound to intervene to maintain the money within the prescribed range, and they can borrow money for the purpose from special funds. In case of lasting imbalance, a readjustment is authorized. (There were ten such realignments since 1979.) Not all members of the EEC belong to the EMS; Britain still remains outside, while Italy enjoys a wider range. Altogether, the fluctuations within the EEC have been reduced, but the goal of joint monetary and financial union is very distant. For a brief summary, see European Community Commission, *The European Monetary System,* the European File no. 15/86 (Brussels: EEC, October 1986).

7. For discussions of the financial crisis, see Pierre Mauroy, *C'est ici le chemin* (Paris: Flammarion, 1982), pp. 18–19; Jean Peyrelevade (who was then on Mauroy's staff), "Témoignage: fallait-il dévaluer en mai 1981?" *Revue politique et parlementaire,* nos. 916–17 (May–June 1985), pp. 128–31; and Philippe Bauchard, *La Guerre des deux roses* (Paris: Grasset, 1986), pp. 26–33.

8. Set up in 1963, the Cour de Sûreté de l'Etat dealt with all cases of infringement against the authority of the state. Its draftsmen had been influenced by the recent struggle against the Algerian liberation movement (FLN) and the Secret Army Organization (OAS). Half the judges were drawn from the civilian judiciary system, and half were military. They were appointed by the government, and the principle of irremovability was waived in their case.

9. It is estimated that, finally, some 132,000 immigrants took advantage of the offer.

10. Initiated by Alain Peyrefitte, then minister of justice, the law designed "to strengthen the security and protect the freedom of persons" was voted on December 19, 1980, and promulgated on February 2, 1981 (*Journal officiel,* February 3, 1981), pp. 415–25. It involved the biggest changes in penal procedure since 1959. Its primary purpose was to speed up punishment and put the emphasis on indemnity for the victims. It reinforced the powers of the police, which was authorized to carry out identity checks in the streets and other public places and allowed to extend provisional arrest to three days. For a critical assessment of

the whole approach by the future Socialist minister of justice, see Robert Badinter, "L'Esprit d'une loi," *Le Monde,* May 10, 1980, p. 1.

11. Robert Badinter, Jacques Delors, and Michel Rocard were among the ministers who favored majority shareholding rather than full nationalization.
12. The Constitutional Council, set up by the 1958 constitution (Articles 56 to 63), supervises the regularity of elections and decides on the constitutional conformity of the laws. It includes nine members, whose terms last for nine years and are not renewable; one third is changed every three years. The presidents of the Republic, of the National Assembly, and of the Senate each appoint three members. Membership was to alter during the years of Socialist rule.
13. The total initial cost of nationalization, of industry and banks combined, was estimated at 47.2 billion francs plus about 3.5 billion francs for subsequent purchases. These and other data on the subject are taken from the excellent analytical summary by Jacques Blanc and Chantal Brulé, *Les Nationalisations françaises en 1982,* Notes et études documentaires, nos. 4721–4722 (Paris: La Documentation Française, 1983). One must also add interest over fifteen years at the prevailing rate. Allowing for the repayment of capital and interest, over 10 billion francs was being repaid each year in the early period. Strangely, nationalization thus contributed to the forthcoming boom on the Paris Bourse.
14. The expression was used by Laurent Fabius on October 8, 1981, during the debate on the nationalization of steel held in the National Assembly. He was referring to the financial scandal over the project to build the Panama Canal, a project presented by Ferdinand de Lesseps, of Suez fame, and backed by financiers and politicians, which collapsed and thus ruined many small investors. The bankruptcy, in 1889, shook the Third Republic and was one of the most notorious scandals of the nineteenth century.
15. The proportion declines to 18.3 percent of employment and 25.9 percent of investment if fuel and power are excluded (Blanc and Brulé, *Les Nationalisations françaises,* p. 52).
16. Alexandre and Delors, *En sortir ou pas,* p. 127.

17. These two proposals were numbers 62 and 60 in Mitterrand's 110 proposals.
18. Much has been written about the Establishment, notably J.-L. Bodiguel and J.-L. Quermonne, *La Haute Fonction publique sous la Ve République* (Paris: Presses Universitaires de France, 1983); R. F. Kuisel, *Capitalism and the State in Modern France* (Cambridge: Cambridge University Press, 1981); Ezra Suleiman, *Elites in French Society* (Princeton, N.J.: Princeton University Press, 1978); and the supplement in *Le Monde aujourd'hui*, April 28–29, 1985, especially Monique Dagnaud and Dominique Mehl, "Allons enfants de la fratrie," pp. iii–iv.
19. Jean-Pierre Chevènement, *Le Pari sur l'intelligence* (Paris: Flammarion, 1985), p. 75, and Pfister, *A Matignon*, pp. 75–79.
20. Colombani, *Portrait du président*, p. 66.
21. The presence on the staff of Régis Debray, the author of *Revolution Within Revolution*, as a technical adviser, provoked some fears in Washington that Paris might contemplate radical plans for Latin America. The fears were groundless. It is enough to read Debray's two major volumes on foreign policy—*La Puissance et les rêves* (Paris: Gallimard, 1984) and *Les Empires contre l'Europe* (Paris: Gallimard, 1985)—to be convinced that the revolutionary days of Régis Debray belong to the past.
22. Pfister, *A Matignon*.
23. This was the title of a book by Albert Thibaudet, *Le République des professeurs* (Paris: Grasset, 1927).
24. According to Catherine Nay, at an Elysean breakfast, on October 21, Pierre Joxe protested, " 'Give these traitors their rights back? Why don't we bring back Pétain's remains to Douaumont and rehabilitate him while we're at it?' 'Yes, why not? We should think about it!' replied Mitterrand icily, not uttering a word more for the rest of the meeting" (*The Black and the Red: François Mitterrand—The Story of an Ambition*, trans. Alan Sheridan [San Diego, Calif.: Harcourt Brace Jovanovich, 1987], p. 59).

6. The Retreat

1. Pierre Mauroy, *C'est ici le chemin* (Paris: Flammarion, 1982), p. 26.
2. "I thought at the time—in June 1982—that a real industrial policy could still be initiated. I certainly made a mistake" (Jean-Pierre Chevènement, *Le Pari sur l'intelligence* [Paris: Flammarion, 1985], p. 95).
3. For more on Robert Hersant, see pages 177–79.
4. In 1983, taxes on personal income accounted for 13 percent of total taxation in France, for 28 percent in Germany and the United Kingdom, for 37 percent in the United States, and for 39 percent in Sweden (*Personal Income Tax Systems Under Changing Economic Conditions* [Paris: OECD, 1986], p. 11).
5. In 1985, the last year of the Socialist government, the wealth tax, levied on property exceeding 3.6 million francs in value, affected only 103,000 taxpayers; the 65 percent income tax rate, about 129,000 taxpayers; and the increased death duties, only a few hundred families (Ministry of the Economy and Finance, *L'Economie française début 1986*, Notes Bleues, no. 265 [Paris, February 1986], p. 19).
6. François Guillaume, leader of the FNSEA during the years when the Socialists were in office, became the minister of agriculture in the government of Jacques Chirac. His predecessor as president of the FNSEA, Michel Debatisse, was a minister under Valéry Giscard d'Estaing.
7. The most notorious example occurred in December 1980, when the Communist mayor of Vitry, in the Paris region, used a bulldozer against a hostel for black immigrants that should have been, in his view, located in another town. But the Communists have since admitted that their line had been wrong, adding that it had also been misinterpreted.
8. During the municipal elections of 1983, Gaston Defferre asked the electors of Marseilles to vote for him, rather than for his right-wing opponent, because he, as minister of the interior, could stop immigration.
9. Pierre Mauroy and his friends have since argued that he had been misunderstood. His point, they claim, was that the new

policies had been introduced almost a year earlier, and by then France was on the right course.
10. Michel Camdessus has since been appointed president of the International Monetary Fund.
11. During the Council of Ministers on February 2, 1983, Mitterrand criticized the interference of the administration with the heads of the nationalized industries. Chevènement took it as a personal attack and later that day wrote a letter of resignation. Nothing happened until the government reshuffle on March 22. Chevènement was then offered other ministries in place of Industry and Research. He refused and resigned. For Chevènement's version, see Chevènement, *Le Pari sur l'intelligence,* pp. 100–101; for the prime minister's version, see Thierry Pfister, *A Matignon au temps de l'union de la gauche* (Paris: Hachette, 1985), pp. 141–42.
12. Among junior ministers appointed two days later and worth mentioning is Huguette Bouchardeau, head of the State Secretariat for Environment and the Quality of Life. Since 1979, Bouchardeau had been the national secretary of the much weakened Unified Socialist party (PSU). When the government was confirming its move to the right, it was worth its while to welcome a newcomer with a leftish reputation, and a junior ministry was not a very high price to pay for Bouchardeau's services.
13. John Vinocur, "Springtime in France: The Socialist Ship Becalmed." *New York Times,* April 26, 1983.
14. Among the sources used to reconstruct the events were Philippe Bauchard, *La Guerre des deux roses* (Paris: Grasset, 1986), pp. 142–53; Jean Marie Colombani, *Portrait du président* (Paris: Gallimard, 1985), pp. 64–65; Serge July, *Les Années Mitterrand* (Paris: Grasset, 1986), pp. 90–101; and Pfister, *A Matignon,* pp. 235–38, 257–73.
15. July, *Les Années Mitterrand,* p. 163.

7. Cultural Counterrevolution

1. People tended to include among foreigners, wrongly, the *harkis*—Algerians who had fought with the French troops

against their country's movement of national liberation and were actually French citizens.
2. The heir to the fortune of the Lambert Cement Company.
3. A useful summary of the National Front is Edwy Plenel and Alain Rollat, *L'Effet Le Pen* (Paris: La Découverte–Le Monde, 1984). For a discussion of the events in Dreux, see Françoise Gaspard and Claude Servan-Schreiber, *La Fin des immigrés* (Paris: Le Seuil, 1984). See also Martin A. Schain, *The National Front in France and the Construction of Political Legitimacy* (London: West European Politics, 1987).
4. For a description, see the lively memoirs of Simone Signoret *La Nostalgie n'est plus ce qu'elle était* (Paris: Editions du Seuil, 1976).
5. For more on the Communist party, see pages 191–95.
6. The vogue was illustrated by the success of two books that glorify the American model: Guy Sorman, *La Révolution conservatrice américaine* (Paris: Fayard, 1983) and *La Solution libérale* (Paris: Fayard, 1984).
7. Alain Geismar, Serge July, and Evlyne Morane, *Vers la guerre civile* (Paris: Editions et Publications Premières, 1969).
8. Franz-Olivier Giesbert, "Les 50 Français les plus riches," *Le Nouvel Observateur,* October 24, 1986, p. 52.
9. The bloody strike at the Talbot car plant at Poissy, near Paris, in the winter of 1983–84 was an illustration of both the general "restructuring" and the specific crisis of labor relations in the car industry. Talbot, like Citroën, by then both owned by Peugeot, had a system based on largely imported labor, tough supervisors, and a puppet union. This system could not survive for long the electoral victory of the Left. When imported slaves begin to talk of dignity, the need for robots becomes particularly imperative. Peugeot reached an agreement with the government over the number of layoffs at Poissy. The CGT reluctantly accepted the deal. The local CFDT aligned itself with the immigrant workers determined to resist. In January, the revolt ended in division between unions and defeat.
10. In the elections to the labor courts on December 8, 1982, the share of the CGT dropped from 42.4 percent to 36.8 percent. In the social-security elections on October 19, 1983, for which comparisons with previous polls cannot be made, the CGT ob-

tained 28.2 percent; the Force Ouvrière, 25.2 percent; and the CFDT, 18.4 percent (*Les Elections à la Sécurité Sociale* in *L'Année politique, économique et sociale en France, 1983* (Paris: Editions du Moniteur, 1984), p. 337.

11. Fortunately, it has not, will be the obvious reply. The Soviets are using radio and television as instruments of propaganda, but in the West, these media are neutral, uncommitted, and nonideological. A simple example will show that the reality is more complex. In Paris, when representatives of workers and employers meet to settle a wage problem or a strike, the radio and television commentators refer to them as "social partners." Were they to say "class opponents," there would be an outcry that they are no longer neutral, are politically motivated, peddle a line, and so on. As a concept, "social partnership" is no less ideological than "class struggle."
12. Jean Riboud, in bed with cancer, was really the inspirer and prime mover of the whole operation.
13. On television and the Berlusconi takeover, see Jean-Michel Quatrepoint, *Histoire secrète des dossiers noirs de la gauche* (Paris: Alain Moreau, 1986). The entire story has a twisted and, for the Socialists, bitter ending. The Chirac government quashed the whole operation, and in the reallocation of channels in early 1987, the Five was awarded to a new team, in which Robert Hersant and Berlusconi were the main partners! It not only had been immoral, but also was not a politically paying proposition.
14. Secondary education became free fifty years later, in 1930.
15. "Through them, the powers of darkness were poisoning our youth," commented *La Croix* in October 1940.
16. In 1871, the Assembly and the Thiers government moved to Versailles, while Paris was run by the Commune. Victorious, the *versaillais* massacred at least 20,000 *communards,* and the name stuck.

8. The Conversion

1. Elie, the duke of Decazes, was described as "leading minister" of Louis XVIII since 1815, when he was thirty-five years old,

but got the official title of "president of the council" only in 1819–20, when he was older than Fabius.
2. The State Council is the supreme administrative jurisdiction in France.
3. For the text of Marchais's report, see Georges Marchais, "Espoir et combat pour l'avenir," *L'Humanité,* February 7, 1985, pp. 3–12.
4. "Mitterrand: J'aime le mouvement qui fait bouger les lignes," interviewed by Serge July, *Libération,* May 10, 1984, pp. 2–7.
5. Quoted in Jacques Attali, "La Mort des mots," *Le Nouvel Observateur,* September 25, 1978, p. 28.
6. Quoted from an address to the alumni of the National School of Administration in "M. Rocard devant les énarques" by Jean-Louis Andréani, *Le Monde,* May 9, 1985, p. 6.
7. Quoted in July, *Les Années Mitterrand,* p. 158.
8. Quoted from a talk with journalists before Mitterrand's television show on January 16, 1985, in ibid., p. 211.
9. Speaking on television in January 1987, Raymond Barre did praise Pierre Bérégovoy, Michel Delebarre, and Jean Auroux for their useful work.
10. The calculation is in Frédéric Bon, "Rétro-simulations proportionnalistes," *Pouvoirs,* no. 32 (Paris: PUF, 1985), pp. 135–47.
11. The population of the 96 *départements* of France (including Corsica) was so divided as to provide 1 deputy per 108,000 inhabitants. But even the smallest *départements* were to have at least 2 deputies, and in 55 the number was rounded off to a higher figure. The Nord was to have 24 and Paris 21 deputies.
12. Their dates of birth are Fabius, 1942; Rocard, 1928; and Mitterrand, 1916.
13. Edwy Plenel and Bertrand Le Gendre.
14. This is suggested in Jacques Derogy and Jean-Marie Pontaut, *Enquête sur trois secrets d'état* (Paris: Robert Laffont, 1986).
15. Jean-Louis Bianco, the secretary general of the Elysée, is reported to have put it plainly (July, *Les Années Mitterrand,* pp. 223–24).
16. Ibid., p. 83.
17. *Verlan* stands for *l'envers,* or reversed slang, as the rhymed cockney of London.

III The European Dimension

1. Interview on French television on December 16, 1984.

9. Reagan's Best Ally

1. In principle, 464 ground-launched Cruise missiles in the German Federal Republic, the United Kingdom, Belgium, the Netherlands, and Italy as well as 108 Pershing-2 missiles only in the Federal Republic.
2. François Mitterrand, *Réflexions sur la politique extérieure de la France* (Paris: Fayard 1986), p. 46.
3. Monica Cossardt and Lutz Bindernagel, "Die Deutsche-Französische Freundschaft hängt doch nicht an einer Tasse Tee," *Stern,* July 9, 1981, pp. 80–84.
4. Mitterrand, *Réflexions,* p. 43.
5. The "Internationale," originally a French song, contains, for instance, the following passage:

    > If these cannibals keep on trying
    > To turn us into heroes
    > They will soon learn that our bullets
    > Are destined for our own generals.

6. The last big public manifestation of this antinuclear movement was probably the mass protest against the Superphénix fast-breeder nuclear reactor at Creys-Malville on July 31, 1977, which turned into a bloody confrontation, with one demonstrator killed and many wounded.
7. Marchais did so first on January 23, 1983, and repeated his endorsement on several occasions. The Communists differed, however, from the governmental line on the inclusion of French (and British) missiles in the calculation of the respective arsenals.
8. What is more, a poll taken by Marplan and published in the *Guardian* on February 16, 1987, shows that if 17 percent do not know and 23 percent approve, the remaining 60 percent of the French people actually disapprove of the presence of American nuclear bases in Europe. This is less than in Germany (66 percent) but more than in Britain (56 percent).

10. In the Footsteps of Predecessors

1. Quoted in François Mitterrand, *Réflexions sur la politique extérieure de la France* (Paris: Fayard, 1986), pp. 318–19.
2. The export of arms was justified in traditional terms: If we do not, somebody else will, and we must save the jobs of French workers. Nobody has yet suggested that Colombian peasants should keep on growing coca for cocaine, yet their living standards are not higher than those of arms producers in France or the United States.
3. Quoted in Mitterrand, *Réflexions,* p. 343.
4. Addressing the United Nations General Assembly, in New York, on September 28, 1983, Mitterrand estimated that the United States and the Soviet Union had about 9,000 nuclear warheads each, compared with France's 98. But the Socialists increased that arsenal. Just the sixth missile launching submarine, the *Inflexible,* doubled that number: It had 16 missiles with 6 warheads each. A seventh such submarine was commissioned by the Socialist government, and older submarines were to be switched to more modern missiles with multiple warheads.

11. Europe and the Nation-State

1. The voting system is already weighted in relation to a country's numerical importance.
2. The term *L'Europe des patries* was coined by Michel Debré, prime minister under Charles de Gaulle from 1958 to 1962.
3. D. Martin, "France: Gaullism, the Left and Foreign Policy," *International Socialist Journal,* no. 22 (August 1967), pp. 597–609.

12. Socialism and National Frontiers

1. In principle, socialism does not tolerate the exploitation of nation by nation any more than that of some people by others.

13. The Case For and Against Mitterrand

1. The international comparisons of gross national product, industrial production, and consumer prices are based on comparable OECD figures, brought up to date and revised, in *Demand and Output, OECD Economic Outlook* 40 (Paris, December 1986), pp. 18–25, 156–57. The figures that back the Socialist case can be checked in Ministry of the Economy and Finance, *L'Economie française début 1986,* Notes Bleues, no. 265 (Paris, February 1986), pp. 5–7, 31–33, 43–46. Those that back the case of the new government are in Ministry of the Economy and Finance, *Rapport économique et financier,* Notes Bleues, no. 302 (Paris, October 1986), pp. 15–23.

    INSEE puts the number of jobs lost between December 1980 and December 1985 at 340,000, if the TUC (provisional and part-time jobs for the young) are included, and at 430,000, if they are excluded. In 1985, the French gross public debt represented only 33.4 percent of the gross national product, compared with 46.6 percent in the United States, 67.2 percent in Japan, and 54.4 percent in the United Kingdom (Jean-Claude Chouraqui, Brian Jones, and Robert Bruce Montador, "Public Debt in a Medium-Term Perspective," *OECD Economic Studies,* no. 7 [Autumn 1986], pp. 103–54). Estimates published by Lloyd's Bank in 1985 put France's gross external debt (minus banking sector assets) at 10.5 percent of its gross national product, compared with 12.7 percent in the German Federal Republic and 15.4 percent in Great Britain ("First World Debt," in *Lloyd's Bank International Financial Outlook* (London, August 1985).

2. Among its achievements may be mentioned social security restitution payments for abortion, a measure that was carried with serious resistance (Claude Estier and Véronique Neiertz, *Véridique histoire d'un septennat peu ordinaire* [Paris: Grasset, 1987], chap. 6).

3. The procedure was again reversed by the government of Jacques Chirac.

4. The number of person-days lost through strikes went down to 731,000 in 1985. The average from 1981 to 1985 was 1.4 mil-

lion, compared with 3.9 million from 1971 to 1975 and 3.2 million from 1976 to 1980. OECD, *Main Economic Indicators, Historical Statistics, 1964–1983* (Paris: OECD, 1984), p. 302; and *Main Economic Indicators* (Paris, June 1987), p. 116.

14. Crisis and Polarization, or For Whom the Bell Tolls

1. François Mitterrand, *The Wheat and the Chaff*, trans. Richard Woodward et al. (New York: Seaver Books/Lattès, 1982), p. 182.
2. Speech at the National Convention of the Socialist party, November 1978 (Mitterrand, "La Ligne d'Epinay," *Unité*, December 8–14, 1978), pp. 10–11.
3. Speech to the Managing Committee of the Socialist party, July 8, 1978, quoted in François Mitterrand, *Politique 2, 1977–1981* (Paris: Fayard, 1981), pp. 205–8.

16. The Negative Lesson

1. Marx develops this point in *Grundrisse der Kritik der politischen Okonomie*. For English text see Karl Marx, *Grundrisse, Introduction to the Critique of Political Economy*, trans. Martin Nicolaus (London: Penguin Books, 1973), particularly the chapter on capital, pp. 704–6.
2. *Tapis* is the French word for "carpet" and is pronounced the same way as Tapie.

Index

Affaire des fuites, 83
Algerian war
  amnesty for, 124
  disagreement over, 124
  participants in, 124
Alternance in French political life, 217–18
American State Department on "Red peril" in Mitterrand's government, 103
Amnesty for Algerian war participants, disagreement over, 124
Attali, Jacques, 266
  as Mitterrand's special adviser, 122
  on Socialism, 196
Auroux laws, 115
Autogestion, nationalization and, 116

Badinter, Robert
  liberal handling of justice by, 262
  as minister of justice, 108–10
Bank, nationalization of, 113–14
Barangé laws, 180
Barre, Raymond
  on eve of 1986 parliamentary election, 213–14
  return to power of, 3
Bérégovoy, Pierre
  as minister of finance, 195, 199
  as minister of social affairs and national solidarity, 134–35, 149
  as secretary general, 122
Berlinguer, Enrico, "historical compromise" of, 48
Berlusconi, Silvio, involvement in funding of television station, 175
Bernstein, Eduard, and German Social Democratic party controversy, 45
Berson, Michel, and nationalization process, 112
Bianco, J. P., as secretary general, 122
Blue-collar workers in labor force, 24
Bouchardeau, Huguette, as minister of environment, 195
Bourse, the
  paralysis of, resulting from Mitterrand's election, 104
  rise in shares during Mitterrand's regime, 266
Bush, George, Vice President, on "Red peril" in Mitterrand's government, 103

Cadres, job increases among, 21–23
Canning, George, 239

Chad, French involvement in, 236–38
Chevènement, Jean-Pierre, 63
    as minister of education, 195, 198
    as minister of industry, 135
    scientific research and, 262
Cheysson, Claude
    on Chad, 237
    as minister of external affairs, 100
Chirac, Jacques
    and debate with Fabius, 210–11
    on eve of 1986 parliamentary election, 213, 214
    on New Caledonia, 239
    return to power of, 3
Cohabitation, Mitterrand and, 212–13, 214–15
Cold war, split between Communists and Socialists and, 47
Collective Utility Work (TUC), unemployment and, 200
Comintern, creation of, 46
Common Market. *See* European Economic Community (EEC)
Common Program, 64
    updating of, Communist-Socialist breach and, 70–71
Communism, international
    Krushchev's "secret speech" and, 37–41
    monolithic, 37–38
    polycentric, evolution of, 40–45
Communist party, French
    attitude of, toward Euromissiles, 229
    blunders of, 193
    departure from Mitterrand's government, 190–91
    general strike in May 1968 and, 58–59, 60
    ministries granted to, by Mitterrand, 102
    in Mitterrand's government, 102–4
    post–Second World War, 60–61
    and Socialists, cooperation between
        in 1960s, 62
        weakening of, 69–72
Communists
    Mitterrand and, 264–65
    and Socialists, split between, and cold war, 47
Coty, René, conversations with Mitterrand, 86
Council of Ministers, under Mitterrand, 121
Cresson, Edith, as minister of agriculture, and FNSEA, 137–38
Cultural counterrevolution, 153–88

d'Estaing, Valéry Giscard
    defeat of, by Mitterrand, 74–75
    as president, election of, 65
    return to power of, 3
de Cuellar, Javier Perez, and Greenpeace scandal, 206
de Gaulle, Charles, 3
    challenge of, 89–90
    on European integration, 249
    on French influence overseas, 239
    meeting of Mitterrand with, in 1943, 80–81
    as nationalist, 249–50
    as president, election of
        in 1958, Communist party and, 61
        in 1965, 58
    return to power of, in 1958, 56–57
    Mitterrand and, 84
    UDR state of, 119
de La Genière, Renaud, as governor of Bank of France, 105
Debré law, 180, 184
Defferre, Gaston, 63
    as minister of interior, and decentralization, 110

INDEX

Delebarre, Michel, as minister of labor, 199
Delors, Jacques
  departure from Mitterrand's government, 195
  as minister of economy and finance, 100, 149
  Second Left and, 167
Democracy
  social
    definition of, 45
    failure of, 275–76
    state and, in "reinvention" of Socialism, 278–79
Democratic and Socialist Union of the Resistance (UDSR), Mitterrand's activities in, 82
Dumas, Roland, on Eureka, 241

Ecology in "reinvention" of Socialism, 282
Economic crisis
  d'Estaing and, 65–66
  ideology and, 273–74
  labor movement and, 53–54
Economic policy of Mitterrand's government, battle over, 144
Education
  battle over, 179–86
  resolution of, 197–98
  of public servants, 120
Election(s)
  local, in 1983, losses by Left in, 145–47
  of Mitterrand, reaction to, 4–7
Electoral reform by Mitterrand's government, 201–2
Elysée, the, as seat of power in Mitterrand's government, 122
Employment expanding sources of, 20–23
  in *cadres,* 21–23
  among employees, 23–25
  in management, 22–23
  in the social services, 20–21
Eurocommunism, evolution of, 41–44
Euroleft, 48
Euromissiles
  Communist party attitude toward, 229
  French attitude toward, reasons for, 227–29
  Mitterrand on, 223, 224–25
Europe
  socialist, potential for, 253–55
  Western, united, controversy over, 247–49
European Economic Community (EEC)
  France and, 33
  Mitterrand and, 242–47
  parliamentary election of, in 1984, 186–87
European Left, political position of, post–Second World War, 51–52
European Research Coordination Agency (Eureka), 241
Executive in Fifth Republic, powers of, 57

Fabius, Laurent
  as minister of industry and research, 149
  on monetary policy, 148
  as prime minister, 189–90, 195–96, 198
  and Chirac, debate with, 210–11
  Collective Utility Work program of, 200
  on Greenpeace scandal, 206, 207
  and Mitterrand, strained relationship with, 211–12
  at Toulouse congress, 209, 210

Falloux law, 180
Farming, changes in, 17–19
Federation of the Democratic and Socialist Left (FGDS), creation of, 62
Feminism, trade unions and, 52–53
Fifth Republic, election procedure in, 62
First World War, international socialism and, 46
Fiterman, Charles, as minister of state in charge of transport, 102
Foreign laborers
  in changing work force, 27–29
  inconsistency of Left on, 142–43
  as scapegoats of opposition, 141–42
  trade union representation of, 52
Foreign trade, growth of, 32–33
Fourth Republic, collapse of, Mitterrand's reaction to, 85–86
Franc, French
  devaluations of, 127–28, 132–34
  fall of, resulting from Mitterrand's election, 104–5
France, changing, in expanding Europe, 15–36
Freedoms, personal, Mitterrand's proposals on, 108–10

Gaspard, Françoise, as mayor of Dreux, 153
General Directorate for External Security (DGSE), in Greenpeace scandal, 204, 208
German reunification, France's fear of, and Mitterrand's foreign policy, 226
German Social Democratic party (SPD), 45–46
Glucksmann, André, on nuclear issue, 230–31

Gouze, Danielle, 80
Greenpeace scandal, 203–8
  Le Monde and, 205
Gulag Archipelago (Solzhenitsyn), French reaction to, 66–67

Habré, Hissène, in Chad, 237
Hernu, Charles
  on Eureka, 241
  as minister of defense in Greenpeace scandal, 204, 205–6, 207, 208
Hersant, Robert, newspaper empire of, 177–78, 179
Hôtel Matignon, as seat of power in Mitterrand's government, 122–23

Immigrant workers
  in changing work force, 27–29
  trade union representation of, 52
Immigrants
  inconsistency of Left on, 142–43
  as scapegoats of opposition, 141–42
Industrial conglomerates, nationalization of, 113
Israel, Mitterrand's visit to, 234–35

Jaruzelski, Wojciech, visit of, 212
Jaurès, Jean, honoring of, by Mitterrand, 98
Jospin, Lionel
  position in Mitterrand's government, 123
  at Toulouse congress, 210
Joxe, Pierre, position in Mitterrand's government, 123

Kanak National Liberation Front (FNLKS), 238–39

INDEX 319

Kautsky, Karl, involvement in German Social Democratic party controversy, 46
Keeper of the Seal, Mitterrand as, 82
Keynesian policy, problems with, 129–30
Khrushchev, Nikita, impact of "secret speech" on international Communist movement, 37–41

European, and economic crisis, 274
Legislation, press, handling of, 176–79
Libération, and ideological shift, 162–63
Living standards, improvements in, 29–31
Luxemburg, Rosa, involvement in German Social Democratic party controversy, 45–46

Labor, Mitterrand and, 265–66
Labor movement
  double failure of, as explanation for ideological shift, 163–65
  in "reinvention" of Socialism, 280–81
Labor unions, social upheaval and, 52–54
Lacaste, Pierre, involvement in Greenpeace scandal, 206
Laignel, André, on role of political minority, 118
Lang, Jack, accomplishments of, as minister of culture, 176, 262
Lange, David, on Greenpeace scandal, 205
Le Monde, and ideological shift, 162
Le Nouvel Observateur, and ideological shift, 163
Le Pen, Jean-Marie
  National Front and, 4
  in European Economic Community parliamentary election of 1984, 187
  political career of, 154–56
Lebanon, French contingent sent to, 235–36
Left, the
  bankruptcy of, 165–76

Mafart, Alain, involvement in Greenpeace scandal, 204, 206–7
Maire, Edmond, as leader of French Democratic Labor Confederation (CFDT), 167–68
Management, job increases in, 22–23
Marchais, Georges
  Communist party under, in Mitterrand's government, 191–93
  on Communist-Socialist cooperation, 69
  on Communist-Socialist split, 72
Mauroy, Pierre, 63
  nationalization and, 112
  as prime minister, 99–100
  resignation of, 188
  response of, to employers, 131
  on withdrawal from European Monetary System, 147–48
Media
  Euromissile issue and, 230–31
  on ideological shift in Mitterrand's government, 162–63
Mendès-France, Pierre
  government of, Mitterrand as minister of the interior in, 82
  Mitterrand compared with, 84

Mermaz, Louis, as president of National Assembly during Mitterrand's government, 123
Mexico City, Mitterrand's speech in, 232–33
Miners, British, 1984–85 strike of, 54–55
Mitterrand, François
  ambition of, 83–84
  attempt on life of, 87–88
  Auroux laws and, 115
  biographical background of, 79
  campaign of, for 1981 election, 72–73
  case for and against, 259–68
    Communists and, 264–65
    ideological retreat and, 266–67
    labor and, 265–66
    nationalization and, 263–64
    social reforms and, 262–63
    statistical, 259–61
  and challenge of de Gaulle, 89–90
  on Communist-Socialist cooperation, 69, 73
  consensus road for France and, 269–71
  conversion of, to Socialism, 86–87
  de Gaulle's return to power in 1958 and, 84
  defeat of, in 1974, and psychological impact, 92–93
  election of, 5
  and Euromissiles, 223–25
  in European Economic Community, 242–47
  Federation of Democratic and Socialist Left and, 62
  first presidential election bid of, 58
  Fourth Republic, collapse of, and, 320
  future of, 294–95
  government of
    administrative changes in, 119–21
    alternance and, 217–18
    ceremonies in, 121
    cohabitation of, with Right, 212–13, 214–15
    Communists in, 191–93
    Communists leaving, 190–91
    electoral reform by, 201–2
    failure of, to present alternative policy, 125–27
    ideological shift in, 161–70
      bankruptcy of Left and, 165–76
      labor movement failure and, 163–65
      media response to, 162–63
    ideological surrender of, 151–52
    intellectual mood during, 161–62
    loss of support for, 135–40
    nuclear issue under, 115
    opposition to, 137–40
      by farmers, 137–38
      by middle class, 138
      press and, 137
      by right-wing students, 138–39
      by truck drivers, 139
    policy of
      in Chad, 236–38
      in New Caledonia, 238–39
      toward Third World nations, 234
    press battle of, 178–79
    press legislation of, 176–79
    proposals of
      for nationalization, 111–15
      for personal freedoms, 108–10
      social, 106–8
      for structure of political and social activity, 110–11
    school battle of, 179–86

Mitterand, François (*Cont.*)
  on Strategic Defense Initiative, 240–41
  unemployment under, solution for, 200
  involvement of, in Greenpeace scandal, 207–8
  as minister of the interior, 82
  as minister of overseas territories, 82
  as minister of state in charge of justice, 82
  on "mixed economy," 196, 197
  National Assembly under, 124
  political background of, 79–83
  as president
    changes in, 215–17
    election of, 76–77
    ceremony following, 97–98
    government appointments of, 100–101
    parliamentary landslide under, 101–2
    speech delivered in Mexico City, 232–33
    travels of, 219–20
    as Reagan ally, 223–31
  in Senate, 88–89
  as Socialist party secretary, 63–64
  underground activities of, 80–81
Mollet, Guy, on Mitterrand as Socialist, 92
Monnet, Jean, as father of European integration, 242
Montand, Yves, political activities of, 157–58
Moscow, visit of Mitterrand to, 234
Moulin, Jean, honoring of, by Mitterrand, 98

National Assembly
  landslide in, under Mitterrand, 102
  under Mitterrand, 124
National Federation of Farmers Unions (FNSEA), 137–38
National Front, success of, 4, 153–54, 156–57
  in European Economic Community parliamentary election of 1984, 187
National frontiers, socialism and, 251–56
Nationalization, 111–16
New Caledonia, French policy on, 238–39
New philosophy
  Gulag Archipelago and, 66–67
  political success of, 68–69
Nouveaux philosophes. *See* New philosophy
Nuclear issue, Mitterrand's pledge concerning, 115

Oueddei, Goukouni, in Chad, 236–37, 238

Palais Bourbon, as seat of power in Mitterrand's government, 123
Parliamentary landslide under Mitterrand, 101–2
Personal freedoms, Mitterrand's proposals on, 108–10
Philosophy, new. *See* New philosophy
Pisani, Edgard, role in New Caledonia negotiations, 238–39
Pleven, René, head of right wing of UDSR, 82
Political activity, structure of, Mitterrand's proposals on, 110–11
Pompidou, Georges
  death of, political situation after, 64–65
  as president
    election of, in 1969, 59

Pompidou, Georges (*Cont.*)
   on eve of 1986 parliamentary election, 213
Poujade, Pierre, 154–55, 156
Poujadism, 154–55
President in Fifth Republic, election of, 57–58
Press, battle over, 178–79
Press legislation, handling of, 170, 176–79
Prieur, Dominique, involvement in Greenpeace scandal, 204, 206–7
Prime minister
   Fabius as, 189–90, 195–96, 198
   Mauroy as, 99–100
   resignation of, 188
Productivity, increased, and rise in unemployment, 273
Professionals, job increases among, 23
Proportional representation, change in electoral system to, 201–2
Protectionism, case against, 252
Public servants, education of, 120

Qaddafi, Muammar, in Chad, 236
Questiaux, Nicole, resignation of, as minister of national solidarity, 134
Quilès, Paul
   involvement in Greenpeace scandal, 206
   on role of Socialists, 118

Radio, Socialists and, 171–74
Reagan, Ronald
   French admiration of, 199
   Mitterrand as ally of, 223–31
"Red peril" in Mitterrand's government, American view of, 103
Researchers, scientific, job increases among, 23

Riboud, Christophe, role in funding of television station, 175
Riboud, Jean, role as personal adviser, 144, 151
Right, the
   extreme, swing of, after postwar consensus, 271–72
   loss of, in 1981, reasons for, 73–74
Rocard, Michel, 266, 267
   as minister of agriculture, 149
   as minister of planning, 100–101
   resignation of, 202–3
   Second Left and, 167
   on Socialism, 196
   as Socialist party leader, 71
   at Toulouse congress, 209–10
Roudy, Yvette, feminist movement and, 262
Rousselet, André, in newspaper negotiations, 178

Sartre, Jean-Paul, on Communist party, 194
Saulnier, Jean, involvement in Greenpeace scandal, 207
Savary, Alain, as minister of education, 182–83
Schoelcher, Victor, honoring of, by Mitterrand, 98
Schools, battle over, 179–86
   resolution of, 197–98
Schuman, Robert, as father of European integration, 242
Scientific researchers, job increases among, 23
Second Left, on Mitterrand, 72
Second World War, international socialism and, 46–47
Service industries
   job increases among, 20–21
   women in, 26–27
Seydoux, Jérôme, role in funding of television station, 175

Signoret, Simone, political activities of, 157–58
Single European Act, 245
Social activity, structure of, Mitterrand's proposals on, 110–11
Social democracy
  definition of, 45
  failure of, 275–76
Social Democratic party, requirements for, 47–48
Social Democratic party (SDP) in Britain, formation of, 47
Social proposals, Mitterrand's 106–8
Socialism
  French experience with, 277–78
  as negative lesson, 286–95
  international
    First World War and, 46
    Second World War and, 46–47
  national frontiers and, 251–56
  "reinvention" of, 277–85
    as alternative to capitalism, 283–85
    ecology in, 282
    labor movement in, 280–81
    state and democracy in, 278–79
    Third World in, 283
    women's liberation in, 281–82
  spread of, in Europe, potential for, 253–55
Socialist party, French
  and Communists
    cooperation between, in 1960s, 62
    weakening of, 69–72
    split between, cold war and, 47
  handicaps of, in 1981, 75–76
  new, Mitterrand as first secretary of, 91
  secretary of, Mitterrand's election as, 63–64

Solidarity, French response to, 159–60
Solzhenitsyn, Aleksandr, as author of *Gulag Archipelago*, 66. See also *Gulag Archipelago*
Spoils system in France, 119
"Star Wars," French opposition to, 240–41
State and democracy in "reinvention" of Socialism, 278–79
Stirbois, Jean-Pierre, as second in command of National Front, 153
Strategic Defense Initiative (SDI), French opposition to, 240–41
Strike, general, in May 1968, 58–59

Taxes, under Mitterrand, 136–37
Teachers, job increases among, 23
Television, government and, 173–76
Thatcher, Margaret, election of, political significance of, 50–51
Third International, creation of, 46
Third World
  Mitterrand's policy toward, 234
  role in "reinvention" of Socialism, 283
Toulouse congress, 209–10
Training of public servants, 120
Tricot, Bernard, involvement in Greenpeace scandal, 205
Turenges, involvement in Greenpeace scandal, 204, 206–7

"UDR state," of de Gaulle, 119
Unemployment
  Mitterrand government's solution for, 200
  rise in, improved productivity and, 273

Unions, labor, social upheaval and, 52–54
Urban growth, postwar, 140–41

Veil, Simone, 157
   involvement in European Economic Community parliamentary election of 1984, 187

Wages, increase in, 29–30
Welfare state in Western Europe, 49–50
White-collar "revolution," foreign labor and, 29
Women in work force, 23–27
   trade union representation of, 52–53
Women's liberation in "reinvention" of Socialism, 281–82
Work force
   changes in, 19–30
      among blue-collar workers, 24
      in employment sources, 20–23
      foreign laborers as, 27–29
      unskilled jobs, 24–25
      wages and, 29–30
      women and, 23–27
   loss of support for Mitterrand's government among, 164–65
World War I. *See* First World War
World War II. *See* Second World War